Thomas Edison

Significant Figures in World History

Leonardo da Vinci: A Reference Guide to His Life and Works,
by Allison Lee Palmer, 2019.

Michelangelo: A Reference Guide to His Life and Works,
by Lilian H. Zirpolo, 2020.

Robert E. Lee: A Reference Guide to His Life and Works,
by James I. Robertson Jr., 2019.

John F. Kennedy: A Reference Guide to His Life and Works,
by Ian James Bickerton, 2019.

Florence Nightingale: A Reference Guide to Her Life and Works,
by Lynn McDonald, 2019.

Napoléon Bonaparte: A Reference Guide to His Life and Works,
by Joshua Meeks, 2019.

Nelson Mandela: A Reference Guide to His Life and Works,
by Aran S. MacKinnon, 2020.

Winston Churchill: A Reference Guide to His Life and Works,
by Christopher Catherwood, 2020.

Catherine the Great: A Reference Guide to Her Life and Works,
by Alexander Kamenskii, 2020.

Golda Meir: A Reference Guide to Her Life and Works,
by Meron Medzini, 2020.

Karl Marx: A Reference Guide to His Life and Works,
by Frank Elwell, Brian Andrews, and Kenneth S. Hicks, 2020.

Eva Perón: A Reference Guide to Her Life and Works,
by María Belén Rabadán Vega and Mirna Vohnsen, 2021.

Adolf Hitler: A Reference Guide to His Life and Works,
by Steven P. Remy, 2021.

Sigmund Freud: A Reference Guide to His Life and Works,
by Alistair Ross, 2022.

Henry VIII: A Reference Guide to His Life and Works,
by Clayton Drees, 2022.

Harriet Tubman: A Reference Guide to Her Life and Works,
by Kate Clifford Larson, 2022.

Joseph Stalin: A Reference Guide to His Life and Works,
by David R. Marples and Alla Hurska, 2022.

Joan of Arc: A Reference Guide to Her Life and Works,
by Scott Manning, 2023.

Charlie Chaplin: A Reference Guide to His Life and Works,
by John W. Fawell, 2023.

Caravaggio: A Reference Guide to His Life and Works,
by Lilian H. Zirpolo, 2023.

Martin Luther King Jr.: A Reference Guide to His Life and Works,
by Peter J. Ling and David Deverick, 2023.

Thomas Edison: A Reference Guide to His Life and Works,
by Paul Israel, 2024.

Thomas Edison

Significant Figures in World History

Thomas Edison

A Reference Guide to His Life and Works

Paul Israel

ROWMAN & LITTLEFIELD
Lanham • Boulder • New York • London

Published by Rowman & Littlefield
An imprint of The Rowman & Littlefield Publishing Group, Inc.
4501 Forbes Boulevard, Suite 200, Lanham, Maryland 20706
www.rowman.com

86-90 Paul Street, London EC2A 4NE, United Kingdom

British Library Cataloguing in Publication Information Available

Library of Congress Cataloging-in-Publication Data

Names: Israel, Paul, author.
Title: Thomas Edison : a reference guide to his life and works /
  Paul Israel.
Description: Lanham : Rowman & Littlefield, [2024] | Series: Significant
  figures in world history | Includes bibliographical references and
  index. | Summary: "Thomas A. Edison: A Reference Guide to His Life and
  Works includes a chronology, entries detailing Edison's inventions,
  laboratories, business enterprises, public image including his relations
  with the scientific community. The bibliography includes his
  publications, biographies and thirteen subject areas related to Edison's
  work and influence" —Provided by publisher.
Identifiers: LCCN 2023048182 (print) | LCCN 2023048183 (ebook) | ISBN
  9781538134269 (cloth) | ISBN 9781538134276 (epub)
Subjects: LCSH: Edison, Thomas A. (Thomas Alva), 1847–1931—Dictionaries. |
  Edison, Thomas A. (Thomas Alva), 1847–1931—Chronology. | Edison, Thomas
  A. (Thomas Alva), 1847–1931—Bibliography. | Inventors—United
  States—Biography.
Classification: LCC TK140.E3 I873 2024 (print) | LCC TK140.E3 (ebook) |
  DDC 621.3092—dc23/eng/20240116
LC record available at https://lccn.loc.gov/2023048182
LC ebook record available at https://lccn.loc.gov/2023048183

Contents

Preface

The lifetime of Thomas Alva Edison (1847–1941) spanned a period of dramatic technological and economic transformations in the United States and Europe—changes at least as disorienting as those wrought by the digital age of our new millennium. Edison was not individually responsible, of course, but he had an outsized role. He was born into a world lighted by flame, such as candles and oil- or gas-burning lamps. Communication most often required physical movement under human or animal power (though the electric telegraph and steam locomotive were already changing the scales of time and distance). Auditory and most visual experiences were ephemeral; once completed, they existed only in memory. The electric light, telegraph, telephone, phonograph (recorded sound), and motion pictures—fields that Edison pioneered or significantly advanced—changed all that.

Edison's successes also changed attitudes about technology itself. At the time of the Civil War, the very *idea* of technological evolution was still new and not yet taken for granted. "Progress," as Americans often called it, was credited to genius—the genius of relatively few individuals and of the nation's exceptional political system. Edison sought to prove otherwise. At his laboratories in Menlo Park and West Orange, New Jersey, he set out to create a stream of new inventions through a process that he could control. In other words, the material future was not something that just happened; he could make it to order. Major corporations like General Electric and the Bell system, absorbing those lessons, set up their own sophisticated research facilities for creating and improving products through deep scientific research.

Edison organized his laboratories to facilitate the processes of inventing and making money from those inventions. He gathered the tools, skills, intellectual resources, and supplies needed to tackle problems of his choosing. The successful test of a new product did not end the creative work. There followed a back-and-forth with Edison's factories, his sales agents, and eventually his customers; designs were revised, new prototypes tested, manufacturing routines adjusted, and sales strategies debated. "Invention" became just one step in a circular process involving research, development, and marketing—the process we now call "innovation."

These conceptual breakthroughs helped to transform the American economy and reshape everyday life in the industrialized world. More than any single invention, they justify Edison's historical significance. But even Edison's stunning originality and fecundity do not quite seem to explain the enduring public and scholarly fascination with the man. The first biographical sketches of "professor" Edison appeared in 1878, when he was just thirty-one years old. A stream of articles, books, and movies about him has ebbed and flowed ever since, including five major book-length biographies in the past thirty years.

One might give various reasons for why successive generations have found so much to say about the man. He embodies myths of exceptionalism, risk-taking, and hard work that Americans like to tell about themselves. He illustrates the reorganization of life around technology and industry. His successes provide a road map to latter-day entrepreneurs (though his failures are probably more instructive). He was a bona fide celebrity, and the seemingly guileless manipulation of his public persona has rippled across the decades.

There is also a sense in which Edison was at the center of things, his activities intersecting with an uncommonly broad swath of American lives and institutions. (Although it must be said that the world in which he moved was overwhelmingly male and white.) An incomplete list of his correspondents number over 25,000. It includes famous figures: politicians, scientists, business captains, and stage luminaries, some of whom have their own entries in this book. It also includes hundreds (probably thousands) of obscure individuals who sought advice, money, approval, or simply Edison's attention, as in the case of a grieving mother whose son suffered industrial poisoning in his employ. Edison did not pigeonhole himself. He embraced overlapping roles as inventor, entrepreneur, industrialist, government councilor, and public sage (among others), all of which embedded him in different social webs. Even his defining role as an inventor depended on his ability to tap into networks outside the laboratory, specifically, bankers, lawyers, journalists, engineers, and scientists. Edison's distinctive technological contributions were shaped, in turn, by those networks, and studying him is a way to query the continuing interplay of technological creativity with economic, social, and political forces in the modern world.

The authors welcomed the invitation from Rowman & Littlefield to create this guide to Edison's life and work. The introduction serves as a short biography, intended to orient the reader and place Edison in historical context. The chronology provides a time frame for the major events of his life. The A–Z dictionary entries sketch the technologies, people, places, and institutions most important to Edison or his legacy. One of that section's distinctive features is the large number of cross-references (indicated by bold type) to other entries. These indicate the high degree to which subjects overlapped in seemingly disparate parts of Edison's life; he was not one to compartmentalize. (The degree of overlap also makes the index an indispensable tool for using this book.) The bibliography provides guidance to the massive bodies of primary and published materials about Edison. Maps were prepared by Michael Siegel of the Rutgers University geography department.

This book distills nearly forty years of research and publishing by the Thomas A. Edison Papers, and the authors gratefully acknowledge the work of past and present colleagues. The project, based at Rutgers University, makes the vast record of Edison's life available and intelligible to the public, students, and researchers. Our core knowledge comes directly from historical documents—those written by, to, or about Edison, his associates, and family members—many of which the project has mounted online (https://edison digital.rutgers.edu) or published as fully annotated transcriptions (https://edison.rutgers .edu/research/book-edition). The project's work is informed by the insights of other scholars and has benefited from collectors and other individuals generously sharing their knowledge. Its major financial and in-kind support comes from the Rutgers University School of Arts and Sciences, the National Endowment for the Humanities, the National Historical and Public Records Commission, the New Jersey Historical Commission, and the Thomas Edison National Historical Park, with additional help from private foundations and individuals.

Acronyms and Abbreviations

Three abbreviations recur throughout the citations to quotations in the encyclopedia entries:

TAE refers to Thomas Edison.

TAEB indicates the Thomas Edison Papers Book Edition (Johns Hopkins University Press), edited by Paul Israel et al. Nine volumes (as of December 2022) present 3,458 transcribed and annotated documents selected from approximately twenty times that number available. References in the text are by volume:page. Volumes are available in hardcover format and can be downloaded electronically through Project Muse of the Johns Hopkins University Press. Fifteen volumes are planned for the edition. See https://edison.rutgers.edu/research /book-edition.

TAED indicates the Thomas Edison Papers Digital Edition hosted at Rutgers University. This searchable online collection is at https://edisondigital .rutgers.edu. It includes (as of December 2022) images of approximately 153,000 documents spanning most of Edison's life. The Edison Papers constantly expands this collection with new documents and features, including a growing number of text-searchable transcriptions. References in the text are to unique document-identifier codes.

Maps

This map shows where Edison lived, worked, or had significant business connections before he moved to Menlo Park, New Jersey, around his twenty-ninth birthday. *Courtesy of Thomas A. Edison Papers.*

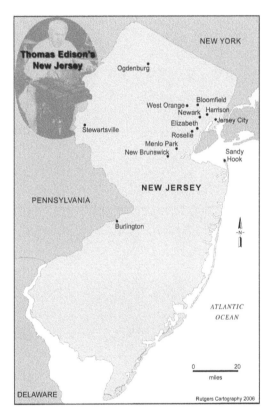

This map details where Edison lived, worked, or did significant business in the state over the course of his career. *Courtesy of Thomas A. Edison Papers.*

Chronology

1847 11 February: Thomas Alva Edison is born in Milan, Ohio, to Nancy (Elliott) and Samuel Ogden Edison as their seventh and last child.

1854 The Edison family moves to Port Huron, Michigan.

1859–1860 Winter: Edison begins selling newspapers and candy on Grand Trunk Railway trains.

1862 Spring: Edison publishes and prints on the train his own newspaper, the *Weekly Herald*. **Fall:** In gratitude for Edison saving the life of his young son from an oncoming freight car, James Mackenzie, the station agent at Mount Clemens, Michigan,, begins to tutor Edison in telegraphy.

1862–1863 Winter: Edison begins working as a telegraph operator in a book and jewelry store in Port Huron.

1863 Spring–summer: Edison works as a telegrapher for the Grand Trunk Railway at Stratford Junction, Ontario, for about six months. He is fired after an alarm he created to awaken him at regular intervals during the night nearly resulted in a train collision.

1863–1864 Edison returns briefly to Port Huron. He works as a railroad telegrapher near Adrian, Michigan, where he meets Ezra Gilliland. He also spends two months as a railroad telegrapher in Fort Wayne, Indiana.

1864–1865 Fall–winter: Edison hones his telegraph skills as an operator for Western Union in Indianapolis, a major railroad hub. He also experiments on improvements in repeaters (devices for retransmitting messages from one line to another).

1865 Spring: Edison moves to Cincinnati; working as a Western Union operator, he also experiments on relays. **17 September:** He becomes a founding member of a Cincinnati chapter of the National Telegraphic Union. **September:** He is promoted to a first-class operator and begins to design devices for multiple telegraphy. **Fall:** Edison moves to Memphis, Tennessee, where he works as a press-wire telegraph operator and experiments with repeaters and duplex systems (for sending two messages simultaneously).

1866 Spring: Edison is employed in Western Union's Louisville, Kentucky, office as a press-wire operator. **1 August:** He leaves for New Orleans, Louisiana, where he intends to embark for Brazil. **Fall:** He returns to work at Western Union in Louisville.

1867 Summer: Edison is employed in the Western Union office in Cincinnati. **October:** He returns to Port Huron.

1868 March–April: Edison moves to Boston as an operator in Western Union's main office. **11 April:** His articles on his telegraph inventions and on the Boston telegraph community appear in *The Telegrapher*. **11 July:**

Edison makes the first of several agreements with Boston businessman E. Baker Welch, who helps finance his early inventive work. **28 July:** Edison signs a patent caveat for a fire alarm telegraph and assigns the invention to Welch. **13 October:** He signs a patent application for an electric vote recorder, resulting in his first patent.

1869 21 January: Edison sells the rights to his first successful printing telegraph (the "Boston instrument") to Joel Hills and William Plummer of Boston. **30 January:** He publicly announces his resignation from Western Union to become a full-time inventor. **Winter–spring:** With Frank Hanaford, Edison establishes a Boston-based business to produce and market private-line telegraphs. **13 April:** His attempt to make his new double transmitter work between Rochester and New York fails. **April–May:** Edison moves to New York City. **22 June:** He receives his first telegraph patent (for the Boston instrument). **ca. 1 August:** Edison succeeds Franklin Pope as superintendent of the Gold and Stock Reporting Telegraph Company, New York City, and improves the company's basic stock printer. **12 September:** He moves to Elizabeth, New Jersey, and boards with Pope's mother. **2 October:** Edison joins with partners Franklin Pope and James Ashley to promote their new electrical engineering and telegraph contracting firm, Pope, Edison & Company. **Fall:** Edison experiments at the electrical instrument factory of Leverett Bradley in Jersey City, New Jersey.

1870 10 February: Edison procures financial support from George Field and Elisha Andrews of the Gold and Stock Telegraph Company to establish a shop and develop inventions. **ca. 15 February:** Edison establishes the Newark Telegraph Works (later Edison and Unger) with William Unger. **1 July:** He becomes a founding partner in the American Printing Telegraph Company. **1 October:** Edison signs a partnership agreement with George Harrington, forming the American Telegraph Works and providing funds for his automatic telegraph experiments. **1 December:** Pope, Edison & Company is dissolved.

1871 Winter–spring: Edison designs typewriters, perforators, ink recorders, and transmitters for automatic telegraphy. **9 April:** Edison's mother, Nancy, dies. **April–May:** Edison manufactures his "cotton instrument" for reporting and printing cotton prices, developed for Gold and Stock. **26 May:** He sells patent rights for existing and future printing telegraphs to Gold and Stock and is appointed the company's consulting engineer. **28–29 July:** Edison begins to keep notebooks to record inventive work. **August:** He manufactures his universal stock printer for Gold and Stock. **October:** He hires Mary Stilwell, his future wife, at his News Reporting Telegraph Company. **25 December:** Edison marries Mary Stilwell.

1872 15–17 January: Edison designs a district telegraph and assigns it to the American District Telegraph Company. **27 January:** He redesigns his universal private-line printer as an electric typewriter for automatic telegraphy. **5 February:** Edison becomes a partner in J. T. Murray and Company (later Edison and Murray). **May:** He delivers the first models of an improved universal private-line printer to Gold and Stock Company. **May–June:** The Exchange Telegraph Company of London begins using his universal stock printer. **3 July:** Edison buys out William Unger's share of their company and dissolves the partnership. **Fall:** Edison tests his automatic telegraph system. **5 November:** Josiah Reiff agrees to pay Edison an annual salary for work on automatic telegraphy. **14 December:** The Automatic Telegraph Company begins using Edison's improvements in automatic telegraphy.

1873 ca. 10 February: Edison verbally agrees with William Orton to develop duplex telegraphy for Western Union. **18 February:** Edison's daughter Marion Estelle ("Dot") is born in Newark. **31 March:** Edison agrees to develop a roman letter telegraph for Harrington and Reiff. **9–22 April:** He drafts ten patent applications for duplex telegraphs. **23 April:** Edison departs for England. **23–27 May:** He conducts tests of his automatic telegraph system for the British Post Office. **ca. 1–15 June:**

He conducts tests of his automatic telegraph system at the Greenwich, England, works of the Telegraph Construction and Maintenance Company. **25 June:** He returns to the United States. **2 August:** Edison drafts a patent caveat containing the basis of the quadruplex telegraph. **25 August:** He executes a patent application for his roman-letter automatic telegraph perforator. **2 September:** Edison sells British rights for his automatic telegraph to a London company. **Summer–Fall:** Edison devises and experiments with circuit designs for duplex, automatic, and cable telegraphy using electromagnets and storage batteries to counter inductive effects. He also executes four patent applications for chemical solutions for telegraph recording paper. **1 October:** Edison demonstrates the quadruplex telegraph for telegraph officials. **28 October:** He executes the first patent caveats for roman-letter automatic telegraph circuits. **Late December:** He conducts tests of his automatic telegraph system for telegraph officials.

1874   27 January: The Automatic Telegraph Company publicly demonstrates Edison's automatic telegraph system. **1 February:** Edison begins experiments with new design for a roman-letter automatic telegraph. **Winter:** He continues experiments on battery and automatic telegraph recording solutions and on cable telegraphy. **March:** He begins developing a district telegraph and fire alarm system. **26 March:** Edison agrees with Joseph Murray and Jarvis Edson to incorporate the Domestic Telegraph Company to exploit his district telegraph and fire alarm inventions. **10 April:** He discovers the electromotograph phenomenon. **ca. 16 April:** Edison demonstrates the roman-letter automatic telegraph to George Ward, American manager of the French and Anglo-American cable lines. **19 May:** Edison seeks the cooperation of Western Union electrician George Prescott in duplex experiments in exchange for a half interest in resulting patents. **22 May:** He begins selling inductorium, an electric shocking device for therapeutic uses. **ca. 1 June:** George Prescott accepts Edison's proposal with approval of Western Union president William Orton; Edison

and Prescott begin quadruplex experiments on Western Union lines and have apparatus made at the Western Union shop. **June:** Edison and Murray prepare instruments for a July demonstration of Edison's fire alarm system in Utica, New York. **Spring:** George Harrington and Josiah Reiff negotiate to form a new company to exploit Edison's automatic telegraph inventions. **8 July:** Edison demonstrates the quadruplex to Western Union officials on the company's lines. **9 July:** He signs his first partnership agreement with George Prescott. **24 July:** Edison finishes several automatic telegraph patent models for George Harrington, including modifications of the British Wheatstone system. **7 August:** He executes a patent application for a telegraph device employing the electromotograph. **19 August:** Edison signs a revised partnership agreement with Prescott and executes several patent applications for duplex and quadruplex telegraphy. **August:** Edison becomes science editor of the *Operator*. **1 September:** Edison publishes first of several articles in the *Operator*. **5 September:** He announces his electromotograph discovery in *Scientific American*. **Late September:** He successfully tests the quadruplex on Western Union line between New York and Boston. **October–November:** Edison installs and modifies the quadruplex on Western Union lines. **ca. 1 November:** Edison moves his family to an apartment on Bank Street, Newark. **ca. 6 November:** He exhibits the electromotograph at the National Academy of Sciences meeting in Philadelphia. **3 December:** He agrees to establish a Domestic Telegraph Company in Canada and begins manufacturing a new private-line printer for the domestic system. **4 December:** Edison successfully establishes quadruplex operation on Western Union line between New York and Chicago with a repeater at Buffalo, New York. **9 December:** The Society of Telegraph Engineers in London elects Edison as a member. **Mid-December:** Edison tries to negotiate sale of the quadruplex to Western Union. **30 December:** Josiah Reiff, John McManus, and Jay Gould reach agreement on the Atlantic and Pacific Telegraph Company (A&P) takeover of Automatic Telegraph Company, with Edison to be the company electrician. **31 December:** Quadruplex patent application suspended for possible interference.

1875 **4 January:** Edison sells his quadruplex patent rights to Jay Gould for $30,000. **5–8 January:** He discusses strategies with Gould for A&P competition with Western Union. **19 January:** William Orton accepts Edison's terms for quadruplex purchase. **23 January:** Edison declines Western Union's terms. **28 January:** Western Union begins legal proceedings against Edison in New Jersey over the quadruplex. **January:** Edison begins serving as electrician for A&P, inspects the New York–Boston line, and begins manufacturing automatic instruments for A&P and domestic telegraph instruments for New York City lines. **24 February:** He executes a patent application with Charles Batchelor for the domestic telegraph system. **3 March:** Edison files two patent applications for a quadruplex repeater and the domestic telegraph system. **March–early April:** Edison moves his family to a house on South Orange Avenue, Newark. **16 April:** He transfers to George Harrington and Josiah Reiff part of his interest in automatic telegraph patents. **20 April:** Edison agrees with Gold and Stock Telegraph Company to settle their outstanding accounts. **30 April:** He begins experiments to develop a new copying process. **7 May:** He begins experiments on an electromotograph repeater for automatic telegraphy. **16 May:** Edison agrees with Joseph Murray to dissolve their manufacturing partnership and sets up a separate laboratory at the Ward Street shop. **31 May:** He prepares a list of proposed experimental topics and conducts his first experiments in his new laboratory. **2 June:** He experiments to find a new force for use in telegraphy. **30 June:** Edison conceives of the electric pen copying system. **26 July:** He proposes the lease of A&P wires for night-letter business to be run from the laboratory. **5 August:** Edison prepares a list of inventions to be displayed at the 1876 Centennial Exhibition in Philadelphia. **September:** He proposes duplexing the Atlantic cable of the Direct United States Cable Company. **Summer:** He develops a basic electric pen copying system

and begins to manufacture and sell it by early September. **23 September:** Edison executes the first caveat for the electric pen copying system. **1 October:** He signs the first of several agreements for electric pen agencies. **ca. 14–21 October:** He develops a new battery for the electric pen. **16 November:** Edison begins acoustic telegraph experiments for Western Union. **22 November:** He executes his first acoustic telegraph caveat and develops new quadruplex designs. He notices a phenomenon he ascribes to an "etheric force" and begins two weeks of intensive experiments on it. **29 November:** Edison gives first newspaper interviews on etheric force. **1 December:** He agrees to assign rights in automatic telegraphy to the new American Automatic Telegraph Company as part of a strategy to force A&P and Jay Gould to settle with those interested in the old Automatic Telegraph Company. **14 December:** Edison signs contracts with Western Union to settle their mutual claims regarding the quadruplex and to formalize the company's support for his work in acoustic telegraphy in exchange for control of his inventions. He conducts etheric force experiments with Professor Henry Morton at the Stevens Institute of Technology in Hoboken, New Jersey. **16 December:** He demonstrates the etheric force with Dr. George Beard at a meeting of the Polytechnic Association of the American Institute of New York. **23 December:** Edison demonstrates the etheric force at a meeting of the Newark Scientific Association. **29 December:** He purchases a house and property in Menlo Park, New Jersey, for his future home and laboratory.

1876 **10 January:** Edison's first son, Thomas Alva Jr. ("Dash"), is born in Newark. **Mid-January:** Edison executes five acoustic telegraph patent caveats. He demonstrates a new quadruplex on Western Union lines and begins constructing a new laboratory at Menlo Park. **7 February:** He improves the design of the electric pen. **13 March:** He files a patent application on his electric pen copying system and arranges the public sale of stock in a new electric pen company. **March–April:** Edison establishes a residence and a new laboratory at Menlo Park. **11 April:** A&P sues Edison and others in New York state court over quadruplex telegraph patent rights. **26 May:** Edison renews his five-year agreement for work with the Gold and Stock Telegraph Company. **June:** Edison's automatic telegraph system and electric pen and automatic press earn awards at the Centennial Exhibition in Philadelphia. **ca. 10 July:** Edison sells British electric pen rights. **24 July:** He conducts etheric force experiments. **25 July:** Edison begins tests of the acoustic transfer telegraph on a line to Philadelphia. **July:** He conducts a sustained series of experiments in telephony. **3 August:** Edison begins two months of experiments on his electromotograph. **August:** He files statements in a patent interference case on acoustic multiple telegraph. **13 September:** He experiments with octruplex acoustic transfer on a telegraph line to Philadelphia. **Summer:** Edison conducts experiments with carbonized paper for electrical resistances, battery carbons, and chemical crucibles. **30 October:** Edison executes patents for a sewing machine motor that uses tuning forks set in motion by electromagnets and for a telegraph recorder/repeater that uses a hole punching method to record messages. **3 November:** He publishes an article in *American Chemist* on the effect of sunlight on various chemicals. **28 November:** Edison agrees to allow Western Electric Manufacturing Company to produce and sell the electric pen and press copying system in the United States. **ca. 1 December:** He and Edward Johnson incorporate the American Novelty Company.

1877 6 January: Edison resolves two patent interference cases with Elisha Gray over acoustic multiple-telegraph design. **Mid-January:** Edison begins new experiments on the etheric force. **20 January:** Edison conducts his first experiments with a telephone transmitter that varies the electrical resistance of a circuit by changing the pressure on carbon. **3 February:** He executes a patent application for an embossing recorder/repeater for telegraphy. **8 February:** Edison conducts his first experiments with a two-plate embossing recorder/repeater. **Mid-February:** He begins

sustained work on the telephone. **14 March:** Edison displays his two-plate embossing recorder/repeater at Western Union's New York headquarters. **18 March:** He devises the first electromotograph telephone receiver. **19 March:** Edison demonstrates his telephone instruments to Western Union officials on a line from New York to Menlo Park. **22 March:** Western Union agrees to provide regular financial support for Edison's Menlo Park laboratory in exchange for all his future telegraph and telephone patents. **23 March:** He begins preparing patent applications for the telephone and a new sextuplex telegraph system. **10 April:** Edison designs a pressure relay and a telephone transmitter that rely on variation in the pressure on semiconductors (chiefly carbon); the transmitter became the basis for his later claim to the invention of the microphone. **24 April:** He signs an agreement with Charles Batchelor, George Bliss, and Charles Holland to market the electric pen in Europe. **26 April–3 May:** Edison testifies in the "Quadruplex Case": *Atlantic & Pacific v. Prescott & others.* **28 April:** Edison begins the exhibition of his "musical" electromotograph telephone at the Newark Opera House. **18 May:** British telegraph engineer William Preece visits Edison at Menlo Park. **31 May:** Edison begins a month of experiments on plumbago mixtures for the telephone transmitter and agrees with George Prescott and Gerritt Smith to pool their British quadruplex patents. **5 June:** Edison loses a patent interference case on acoustic multiple telegraphy to Elisha Gray. **ca. 6 June:** He designs a combination telephone transmitter/electromotograph receiver. **ca. 8 June:** Having completed extensive experiments in the laboratory, he tests his sextuplex telegraph system on the Western Union line from New York to Boston but then abandons it. **3 July:** Edison demonstrates his embossing recorder/repeater to officials of Western Union and the British Post Office. **17–18 July:** He conceives and experiments on a device to record and repeat telephone messages that marks the beginning of the phonograph. **19 July:** Edison's musical telephone is exhibited before a large crowd at the Permanent Exhibition Hall in Philadelphia. **ca. 19 July:** Edison finishes the preliminary application for a British telephone patent that includes a description of the phonograph. **30 July:** Edison begins the development of carbon "fluff" for telephone transmitters. **3 August:** He begins a monthlong correspondence on spectroscopy with astronomer Henry Draper. **4–10 August:** Edison and Thomas David conduct telephone tests between Menlo Park and Manhattan. **10 August:** Edison exhibits to Western Union officials a telephone transmitter/electromotograph receiver built by Joseph Murray. **Mid-August:** He tests various telephone designs on lines in New York City. **20 August:** He demonstrates a new telephone design to Western Union president William Orton and others. **24 August:** He begins design of handheld telephones. **9 September:** Edison resigns his electrician position with the A&P; he conducts arc light experiments. **ca. 10 September:** Edison conducts his first incandescent lighting experiments. **11–13 September:** He goes fishing in Raritan Bay with laboratory staff. **15 September:** Edison signs an agreement with Franklin Badger to market the telephone in Canada. **17 September:** He demonstrates a new transmitter to William Orton, who orders 150 sets of transmitters and receivers. **18 September:** With George Prescott, he signs an agreement with Stephen Field and Cornelius Herz regarding European quadruplex patents. **28 September:** Edison learns of the success of quadruplex telegraph tests made by Gerritt Smith and George Hamilton on British Post Office lines. **29 September:** He learns of the suspension of his first telephone patent application pending the resolution of interference conflicts. **5–8 October:** Edison redesigns the telephone transmitter manufactured by Murray. **18 October:** He participates in Edward Johnson's telephone exhibition in Jersey City. **22 October:** Edison begins using disks of a plumbago–rubber mixture to replace "fluff" in telephone transmitters. **26 October:** He begins using disks of lampblack and rubber in telephone transmitters. **Mid-November:** Edison designs an induction coil circuit for telephones. **November:** He explores alternatives to carbon for telephone transmitters. **1–3 December:** Edison tests the induction coil telephone circuit. **1–6 December:** At Edison's direction, John Kruesi makes the first tinfoil-cylinder phonograph. **7 December:** Edison demonstrates the cylinder phonograph at the New York office of *Scientific American.* **12 December:** He prevails in electric pen patent interference against Edward Stewart. **15 December:** Edison executes the first U.S. phonograph patent application (U.S. Patent 200,521). **17 December:** He signs agreements with George Bliss and Theodore Puskas regarding the sale of his European telephone and phonograph patents. **27 December:** Edison confers with Edward Johnson, Uriah Painter, and Gardiner Hubbard on their proposal to break his Western Union contract and allow them to form a company to market the phonograph. **ca. 29 December:** He is proposed as a scientific member of the American commission to the 1878 Paris Universal Exposition. **31 December:** Edison demonstrates the phonograph to William Orton.

1878 **1–3 January:** Edison exhibits the phonograph at Western Union offices in New York. **8–10 January:** He tests his telephone at Western Union's New York offices, further straining relations with company electricians. **29–30 January:** Edison visits the Ansonia (Connecticut) Clock Company and experiments with applying the phonograph to clocks. **30 January:** He signs an agreement for commercial exploitation of the phonograph in the United States. **January:** He makes flywheel phonographs to give to prominent scientific leaders. **4 February:** Edison begins testing the carbon telephone transmitter between New York and Philadelphia. **ca. 6 February:** He completes the design of a small demonstration phonograph that will eventually be sold at the Paris Universal Exposition. **22 February:** He plans a series of experiments to develop a telephone receiver that does not infringe on Bell patents. **28 February:** Edison executes phonograph and telephone caveats and three related patent applications. **20 March:** The first public demonstration of the improved carbon telephone transmitter takes place at the Franklin Institute, Philadelphia. **22 March:** Edison signs an agreement with the London Stereoscopic and Photographic Company for commercial exploitation of the phonograph in Great Britain. **29 March:** *New York World* publishes first description of Edison's medicinal preparation for relieving pain ("polyform"). **March:** Newspapers begin extensive coverage of Edison, helping to spread his fame and produce crowds of visitors to Menlo Park. **10 April:** The *New York Daily Graphic* dubs Edison the "Wizard of Menlo Park." **18–19 April:** He demonstrates the phonograph for the National Academy of Sciences, members of Congress, and President Rutherford B. Hayes in Washington, D.C. **24 April:** The Edison Speaking Phonograph Company is incorporated. **April:** Edison publishes "The Future of the Phonograph," ghostwritten by Edward Johnson, in the *North American Review.* **1 May:** The Universal Exposition, with Edison's exhibit, opens in Paris. **ca. 8 May:** Edison hires John Ott to develop a toy phonograph. **16 May:** Edison designs a new form of tasimeter, an original ultrasensitive heat-measuring instrument. **May–June:** He arranges phonograph and telephone sales agencies for Australia and Central and South America. **1 June:** Edison secures George Gouraud as his agent for the telephone in Great Britain. **26 June:** He receives his first honorary doctorate, from Union College in Schenectady, New York. **Spring:** Edison attempts to develop a hearing aid and other acoustic devices. **13 July:** He leaves with George Barker for a solar eclipse expedition to Rawlins, Wyoming, and a monthlong vacation in the western United States. **1–3 August:** Edison visits San Francisco. **ca. 6–7 August:** He tours Yosemite. **9 August:** He spends the night at Virginia City, Nevada, where he inspects mines and discusses problems of heat and ventilation in the shafts. **22 August:** Edison receives a Grand Prize from the Paris Universal Exposition. **23 August:** He is introduced as a new member of American Association for the Advancement of Science (AAAS) at its annual meeting in St. Louis, Missouri, and delivers a paper on the tasimeter. **27 August:** He begins electric light experiments. **10 September:** Edison drafts his first patent caveat in electric lighting. **13–16 September:** He announces that he has solved the problem of incandescent electric lighting. **19 September:** Edison begins to receive offers from prospective electric lighting investors. **5 October:** He executes his first patent application for electric lighting. **5–18 October:** Edison works on a new phonograph for business dictation. **16 October:** The Edison Electric Light Company is incorporated, principally by investors connected with Western Union and Drexel, Morgan & Company. **23–26 October:** Edison is confined to bed by facial neuralgia. **26 October:** Edison's third child, William Leslie, is born in Menlo Park. **October:** Begins construction of a new machine shop and office buildings at Menlo Park. **15 November:** He assigns lighting patents to the Edison Electric Light Company for $30,000 plus stock and royalties. **23 November:** Edison endorses Uriah Painter and Edward Johnson to take control of the Edison Speaking Phonograph Company. **29–30 November:** Edison designs the first electric meter. **November–December:** Edison displays various electric lighting devices at Menlo Park for Edison Electric Light Company investors. He compares operating costs of gas, arc, and incandescent lighting systems. **3 December:** He executes his first patent application for an electric generator based on the earlier tuning fork motor. **December:** He negotiates through Grosvenor Lowrey with Drexel, Morgan & Company for foreign rights to electric lighting inventions.

1879 **2 January:** The laboratory machine shop at Menlo Park begins constructing Edison's first original dynamo machine (generator). **19 January:** Edison begins studying the properties of heated wires, experiments that will lead to his discovery of occluded gases. **23 January:** He conducts vacuum lamp experiments with a hand pump. **6 February:** Edison begins several days of experiments on the vacuum lamp using the hand pump. **9 February:** He drafts a caveat on the use of a vacuum to overcome occluded gases in lamp wires. **13 February:** Edison tests his first dynamo machine. **13 March:** Edison begins designing a bipolar dynamo with large field magnets. **ca. 18 March:** Edison demonstrates parallel-wired incandescent platinum lamps to prospective investors. **19 March:** Edison begins planning for the electric illumination of the USS *Jeannette,* commissioned to make an Arctic expedition for the *New York Herald*. **25 March:** Edison begins collecting statistics on gas companies in New York. **26 March:** He acquires a Geissler mercury vacuum pump. **8 April:** Edison begins experiments on generating thermoelectricity. **12 April:** He decides to incorporate a new coil and commutator arrangement into the bipolar dynamo. **29 April:** Edison signs an agreement to sell telephone patents and permit the formation of a telephone company in continental Europe. **ca. 30 April:** He begins testing the first large dynamo. **April:** Edison designs a combination Sprengel and Geissler vacuum pump. **ca. 1 May:** Edison begins drafting a British provisional patent specification for his system of electric lighting and begins a search for platinum deposits in gold mining regions of the United States and Canada. **2 May:** He signs an agreement for the sale of his electric light patents and the formation of a company for continental Europe. **12 May:** He agrees to assign Western Union the rights to the electromotograph telephone receiver. **14 May:** The Edison Telephone Company of Europe, Ltd, is incorporated in New York. **15 May:** Edison signs an agreement to form Edison telephone companies in European countries. **16 May:** He grants George Gouraud half the proceeds from the sale of British patent rights on the electromotograph receiver. **18 May:** Edison makes drawings for an electric railroad. **20 May:** He ships dynamo and other electric lighting equipment to the USS *Jeannette*. **May:** Edison agrees to the formation of an Edison telephone company in England; he begins developing a switchboard for telephone exchanges. **12 June:** He begins an extensive series of tests on chalk buttons for the electromotograph receiver. **9 July:** Edison sends the first shipment of telephones and a switchboard to London. **July:** Edison reconfigures the electromotograph receiver and modifies the carbon transmitter in telephones for Britain. With the chemical staff, he assays ores sent to the laboratory and orders the shop to make apparatus for separating and concentrating ores. **August:** Edison orders Francis Upton to begin analysis of long-distance electric power transmission for mines near Virginia City. **14**

August: U.S. Patent Office declares six interference cases over carbon telephone transmitters. **ca. 20 August:** Edison hires glassblower Ludwig Böhm. **ca. 28 August:** Edison writes a paper on electric light experiments to be presented by Francis Upton at a meeting of the AAAS in Saratoga, New York. **30 August:** Edison demonstrates the telephone at the AAAS meeting. **8 September:** He solicits U.S. manufacturers to loan him sewing machines for electric motor experiments. **11 September:** He sends Frank McLaughlin to California to investigate platinum-bearing sands. **20 September:** Edison files a provisional patent specification in Britain for the telephone switchboard. **September:** He assigns U.S. rights for polyform to Charles Lewis and associates, who organize the Menlo Park Manufacturing Company to market and sell Edison's patent medicine. **1 October:** Western Union orders 100 electromotograph receivers. **3 October:** Edison authorizes George Gouraud to organize telephone companies in British colonies. **7 October:** Edison begins electric light experiments with molded carbon spirals. **14 October:** He signs an agreement with José Husbands to establish a telephone company in Chile; an agreement to combine the French Edison, Gower, and Blake telephone companies is signed in Paris. **15–16 October:** Edison begins experiments to solve problems with chalks for the electromotograph receiver. **22 October:** He conducts the first successful carbon filament lamp experiments with carbonized thread in a vacuum. **1 November:** He executes a patent application for a high-resistance carbon filament lamp. **November:** Francis Upton begins writing an article about Edison's electric light system to be published under Edison's byline in the February 1880 *Scribner's Monthly*. **9 December:** Edison organizes the Edison Ore Milling Company. **11 December:** He applies for a patent covering the manufacture of horseshoe-shaped cardboard lamp filaments. **ca. 29 December:** Edison begins the public exhibition of his electric lighting system in and around the Menlo Park laboratory.

1880 2 January: Edison closes the laboratory to the public after being overwhelmed by visitors; he begins experiments with a high-speed dynamo and low-resistance armature constructed of iron plates. **17 January:** Edison agrees to give exclusive rights to his Australian telephone patents to Western Electric. He accepts a reduced royalty for sales of the electric pen apparatus and sends telephone receivers operated by electric motors to the London *Times*. **27 January:** Edison authorizes Edward Johnson to negotiate royalties with the Edison Telephone Company of London; the proposed consolidation of Edison, Gower, and Blake telephone interests in Paris collapses. **28 January:** Edison executes a patent application for his electrical distribution system. **January:** He hires engineers Charles Clarke and Julius Hornig to help with electric lighting. **12 February:** Edison orders Francis Upton to prepare a preliminary estimate for a central station to power 10,000 lights; Edison and Charles Batchelor begin two weeks of experiments on paper and fiber filaments for lamps. **13 February:** Edison begins experiments to prevent the "electrical carrying" that darkened lamp globes, leading to the discovery of the Edison Effect. **10 March:** He executes a patent for an electric meter. **10–12 March:** Physicists George Barker and Henry Rowland visit Menlo Park to test the thermal efficiency of the lamp. **14 March:** Edison orders Charles Mott to begin keeping a journal of activities at the laboratory. **19 March:** Professors Cyrus Brackett and Charles Young test the efficiency of the dynamo. **25 March:** Edison begins experiments on a magnetic iron-ore separator. **Winter–spring:** Edison and laboratory staff test a wide variety of paper, wood, and vegetable fibers as lamp filament materials. **1 April:** Edison, Francis Jehl, and Otto Moses begin experiments with a simplified single vacuum pump. **3 April:** Edison executes a patent application for a magnetic ore separator; he sells the U.S. patent rights on the motograph relay to Western Union. **23 April:** Edison's assistants begin installing the electric lighting system on the steamship *Columbia*. **24 April:** Edison purchases the former electric pen factory across the railroad tracks at Menlo Park for his lamp factory. **26 April:** The president of Chile grants him an exclusive eight-year license for telephones. **27 April:** Edison and his wife attend a reception aboard the *Columbia* in New York. **April:** Construction on the electric railroad at Menlo Park begins; Edison experiments with frosted lamp bulbs. **13 May:** The electric railroad commences operation. **ca. 13 May:** Edison agrees to finance the publication of *Science*. **1 June:** He agrees to consolidation of the British Edison and Bell Telephone companies into the United Telephone Company, Ltd. **16 June:** Seeking the cause of electrical carrying, Edison and astronomer Charles Young conduct a spectroscopic examination of lamps. **21 June:** He receives a visit from a representative of the Corning Glass Works, which will manufacture the bulbs for Edison lamps. **4 July:** Edison offers to file a dynamo patent application for physicist Henry Rowland as a challenge to an application from the German company Siemans & Halske. **7–19 July:** Edison and Charles Batchelor experiment on an electric-powered air balloon. **9 July:** Edison begins experiments with bamboo filaments. **24 July:** Edison begins experiments with depositing volatile hydrocarbons on lamp filaments. **28 July:** He executes a patent application for a lamp filament of carbonized bamboo or similar fiber; he also executes a patent application with Charles Batchelor for a device to test carbonized lamp filaments. **ca. 30 July:** Edison orders Otto Moses to research published literature on carbonaceous materials and processes of carbonization. **31 July:** Edison executes a patent application for a symmetrical system of electrical distribution to maintain uniform voltage. **July:** He begins collecting statistics on gas light and power usage in a proposed central station power district in Lower Manhattan. **4 August:** Edison executes a patent application for a "feeder and main" system of electrical distribution. **ca. 9 August:** He executes a patent application for a direct-connected steam dynamo. **16–24 August:** He experiments on the vacuum preservation of fruit. **18 August:** Edison signs agreements authorizing the sale of patents for telephone, electric light and power, and electric railways in dozens of foreign countries and colonies. **19 August:** He establishes a line of credit with Drexel, Morgan & Company to finance electric railway experiments.

27 August: Edison sends John Segredor to Georgia, Florida, and Cuba to gather plant samples for possible filament use. **ca. 30 August:** Edison sends inquiries about bamboo to Brazil, Panama, Cuba, and other Caribbean islands. **ca. 31 August:** He designates laboratory assistant Wilson Howell to develop a better insulating compound for underground cables. **28 September:** Edison and Charles Clarke conduct experiments on heating copper rods revolved through magnetic lines of force. **September:** The *North American Review* publishes an article in Edison's name (drafted by Francis Upton) on electric light and power. **ca. 6 October:** Edison sends William Moore to China to seek the best bamboo variety for lamp filaments. **21 October:** Edison and Francis Jehl begin experiments on treating lamp filaments with hydrocarbon vapor. **October:** Edison has Upton make a new estimate of the cost of a 10,000-light central station plant. **8–12 November:** Edison testifies in the telephone interference cases. **ca. 18 November:** He forms the Edison Electric Lamp Company with Batchelor, Upton, and Edward Johnson. **28 November:** Edison and assistant William Hammer begin two weeks of experiments on electrical carrying in lamps. **1 December:** Edison decides to use bamboo filaments in lamps made at his factory. **2 December:** He requests the Edison Electric Light Company to pay electrical engineer Hermann Claudius to map out the feeder and main system of conductors for the first central station district in Lower Manhattan. **11 December:** Edison executes a patent application for an improved dynamo armature using copper bars in the induction circuit. **15 December:** He writes William Crookes about manufacturing the Crookes radiometer at the lamp factory. **17 December:** The Edison Electric Illuminating Company of New York is incorporated. **20 December:** Edison demonstrates his electric light and power system at Menlo Park to New York aldermen and other city officials.

1881 15 January: Edison asks New York lawyer Sherburne Eaton to take over management of the Edison Electric Light Company. **7 February:** He receives an order from Drexel, Morgan & Company to construct a direct-connected steam dynamo for a London exhibition. **ca. 10 February:** Edison agrees to provide equipment for an electric lighting exposition in South America. **26 February:** Edison conducts the first test of the big direct-connected steam dynamo. **28 February:** Samuel Insull becomes Edison's private secretary. **February:** Edison leases office space at 65 Fifth Avenue in New York City, together with the Edison Electric Light Company. **ca. 1 March:** He acquires a shop in New York City at 104 Goerck Street for the Edison Machine Works to manufacture heavy electrical equipment. **4 March:** He incorporates the Edison Tube Works in New York. **8 March:** He signs an agreement for the Edison Lamp Company to supply incandescent lamps to the Edison Electric Light Company. **30 April:** Edison has his staff conduct experiments on preserving fresh fruit in a vacuum. **9 May:** Edison purchases a building in Harrison, New Jersey, for an enlarged lamp factory. **17 May–25 June:** Edison executes twenty-three patent applications on electric lighting. **Spring:** His crews begin laying conductors for the Pearl Street central station in New York City. **May:** He constructs magnetic ore separator for recovering iron ore. **June:** He travels to Quogue, Long Island, to inspect the operation of an iron-ore separator. **July 1:** Edison sends Charles Batchelor to Paris to supervise the Edison exhibit at the International Electrical Exposition and to oversee Edison's electric lighting interests in Europe. **July 15:** He discharges most of the remaining employees at the Menlo Park laboratory. **26 July:** He executes a patent application with Patrick Kenny for a facsimile telegraph. **10 August 10:** Edison's exhibit opens at the Paris Electrical Exposition. **23 August:** Edison approves terms of consolidated control of telephone patents in Europe with other inventors. **28 August:** He travels to Michigan to meet Mary Edison. **August:** He directs Charles Hughes to conduct a preliminary survey for a new electric railroad line at Menlo Park. **7 September:** Edison returns from his trip to Michigan. **14 September:** Edison signs a contract with Henry Villard, who agrees to provide funds for experiments on electric railroads. **September:**

Edison's ore separator is used by the Edison Ore Milling Company to separate iron ore from black sand at Quonocontaug, Rhode Island. **6 November:** Edison agrees to establish electric light and manufacturing companies in Europe. **Fall:** Edison establishes a testing department at the Machine Works that also becomes the headquarters for dynamo and other electric lighting experiments.

1882   12 January: Edison's central station on Holborn Viaduct in London begins operation. **17 January:** Edison's exhibit opens at the Crystal Palace Exhibition in London. **1 March:** Edison travels to Florida to join his family on vacation. **28 March:** He returns to Menlo Park and begins an extensive series of electric lighting experiments at the Menlo Park laboratory. **Spring–summer:** Edison executes fifty-three patent applications covering electric lighting, electric railways, and secondary batteries; he designs a high-voltage direct current distribution system. **July:** Edison leaves on a sailing trip and reaches Montreal by mid-month. **18 July:** He returns to Menlo Park. **July:** Edison designs three-wire electrical distribution system. He learns of the delay or loss of dozens of patent applications due to the malfeasance of attorney Zenas Wilber. **4 September:** He opens the Pearl Street central station in the Wall Street district of New York. **1 October:** Edison leases a house at 25 Gramercy Park South, New York City. **4 October–28 November:** Edison executes thirty-four patent applications covering electric lighting and electric railways. **October:** He closes his Menlo Park laboratory and establishes a laboratory on the top floor of the Bergmann and Company factory in New York City.

1883   19 January: Edison's demonstration of the village-plant electric lighting system begins operating in Roselle, New Jersey. **15 February:** Edison travels to Boston to begin the establishment of an illuminating company. **19–27 February:** Vacations in the South. **29 March:** Edison decides to form a central station Construction Department with Samuel Insull and Edward Johnson. **4–5 April:** Edison travels to Albany and then to the electrical exhibition in Cornwall, Ontario. **16 April:** He conducts a tour for former Mexican president Porfirio Díaz of the Edison lighting plants and manufacturing shops in New York City. **20 April:** He agrees in principle to combine his electric railroad patents with those of Stephen Field in a new company. **3 May:** Edison gives Samuel Insull power of attorney to conduct all business related to the Construction Department. **3–5 May:** Edison goes to Shamokin, Pennsylvania, to sign a contract to construct the first village central station plant in Sunbury, Pennsylvania. **15–18 May:** He asks electric light agents for information on the use of the Maxim light in their regions. **1 June:** He dismisses all domestic servants from his home in New York City. **ca. 1 June:** He hires engineer Frank J. Sprague. **28 June:** Central station service begins in Milan, Italy. **June–12 July:** In Sunbury, Edison supervises construction and opening of the first three-wire village plant. **4 July:** The Sunbury station begins service. **23 July:** Edison donates to the telegraphers' strike relief fund. **24 July:** He conditionally approves a merger of his British electric light interests with those of Joseph Swan. **1 August:** In a New York *Evening Post* interview, Edison claims he will become "simply a businessman for a year." **Early August:** He spends two days on vacation with his family in Long Beach, Long Island. **Summer:** He adopts Frank Sprague's method of determining conductors for central stations. **18–26 September:** Edison oversees the completion of the Shamokin central station. **1 October:** Edison inaugurates Brockton central station. **ca. 1 October:** Edison family moves to the Clarendon Hotel, New York City. **ca. 13 October:** Edison devises a new voltage regulator based on the "Edison Effect" lamp. **13 November:** He retains attorney John Tomlinson for a lamp patent infringement lawsuit. **14 November?** Edison testifies as a plaintiff's witness against overhead electric lines in New York City. **11 December:** He repays more than $35,000 borrowed from Drexel, Morgan & Company. **17–ca. 18 December:** Edison travels to Fall River, Massachusetts, to inaugurate central station service there. **ca. 24 December:** He begins experiments with gelatinous materials for lamp filaments.

1884 24 January: The German patent office declares that all carbon-filament incandescent lamps are subject to Edison's patent. **January:** Edison directs research on the electrodeposition of metallic films; he incorporates the Edison Lamp Company and the Edison Machine Works. **ca. 5 February:** He drafts a contract for central station engineers. **10 February:** He and his wife Mary leave for vacation in Florida. **ca. 23 February:** The Swedish patent office voids a number of Edison patents for lack of use in the country. **February:** Edison withdraws the "Edison Effect" voltage indicator; he loses the retrial of a civil lawsuit brought by Lucy Seyfert. **February–March:** On vacation in Florida, he fills a pocket notebook with experimental plans. **24 March:** Edison receives a copy of the "Dot and Dash" polka, named for Edison's children, written by John Milliken. **31 March:** He and Mary return to New York. **ca. 22–26 April:** Due to illness, Edison cancels a trip to Newburgh, New York. **24 April:** Edison announces a plan to disband the Construction Department. **April–May:** He lobbies for a bill in the New Jersey legislature authorizing electric lighting companies to put up poles or lay wires beneath the streets. **1 May:** The Edisons move back to the Gramercy Park house in New York City. **1–2 May:** Edison demonstrates his electric lighting system at a public exhibition in Worcester, Massachusetts. **7 May:** He accepts the nomination as a vice president of the new American Institute of Electrical Engineers. **ca. 15 May:** Edison prepares a patent caveat on the use of an electromagnet to separate nonferrous ores and instructs John Ott to conduct experiments. **ca. 21 May:** He appoints Charles Batchelor general superintendent of the Machine Works. **28 May:** Edison retains Richard Dyer and Henry Seely as patent attorneys. **22 July:** Edison gives up office space at 65 Fifth Avenue, New York City; he begins a week of concentrated experiments on the direct conversion of coal into electricity. A sheriff's sale of the Menlo Park property takes place to satisfy judgment in the Lucy Seyfert lawsuit. **9 August:** Mary Edison dies at Menlo Park. **23 August:** Edison accepts an appointment as a member of the National Conference of Electricians. **1 September:** He signs stockholder agreements with the manufacturing shops to transfer the Construction Department's business to the Edison Company for Isolated Lighting; he rents the top floor of a residence at 39 East 18th Street in New York City. **4–6, 16–19 September:** He attends the International Electrical Exhibition in Philadelphia with daughter Marion. **ca. 15 September:** Edison ends the lease on the Gramercy Park residence. **24 September:** Edison returns to Philadelphia for several days. **21 October:** A second sheriff's sale of Edison's Menlo Park properties takes place to satisfy judgment in the Seyfert case. **28 October:** Edison Electric stockholders elect a slate of directors that gives Edison greater influence over the company. **October:** He negotiates an arrangement with the American Bell Telephone Company regarding experiments he is conducting on improved telephone technology. **10 November:** Edison attends the Sullivan–Laflin fight at Madison Square Garden. **ca. 20–24 December:** Edison visits Boston to confer with American Bell Telephone. **23 December:** He attends the Boston Electrical Exhibition. **November–December:** Edison designs special lighting for use at Steele MacKaye's Lyceum Theater (New York) and the Bijou Theatre (Boston).

1885 ca. 2–6 January: Edison is in Boston to negotiate a formal contract with American Bell Telephone for his work on long-distance telephony and to revisit the Electrical Exhibition. **ca. 23–26 January:** Edison again visits Boston and probably briefly meets Mina Miller at Ezra and Lillian Gilliland's home. **20 February:** Edison embarks on extended trip with his daughter Marion and Ezra and Lillian Gilliland. **ca. 23 February:** Edison agrees with Ezra Gilliland to transfer joint patent rights for railway telegraphy to the Railway Telegraph and Telephone Company. **ca. 28 February:** Edison's traveling party visits the World's Industrial and Cotton Centennial Exposition in New Orleans. **ca. 5–21 March:** The Edison party vacations in Florida; Edison and Ezra travel to the Gulf Coast and contract to buy land in Fort Myers, Florida. **Winter–spring:** Edison conducts railway telegraph experiments. **3 June:** He dispatches Francis Upton to Europe as his representative for the electric light business. **2–ca. 9 June:** Edison visits Ezra Gilliland in Boston. **27 June–ca. 8 July:** Edison's visit to Gilliland's vacation cottage at Winthrop, Massachusetts, overlaps with the visit of Mina Miller. **14–ca. 22 July:** Edison has an extended stay at Gilliland's vacation cottage at Winthrop. **10–18 August:** Edison visits the Miller family at the Chautauqua Institution. **18–ca. 31 August:** He travels with Marion, Mina, and others to Niagara Falls, Montreal, and New Hampshire; he proposes marriage to Mina. **4 September:** He authorizes the settlement of major outstanding debts due him for the construction of the central station generating plants. **19 September:** Edison sends Alfred Tate to set up a commercial trial of the phonoplex system in Canada. **21–30 September:** Edison visits the Gillilands in Boston. **28 September:** The Edison household moves to the Normandie Hotel in New York. **30 September:** Edison asks Mina Miller's father for permission to marry. **ca. 8 November:** He incorporates the International Railway Telegraph and Telephone Company with other colleagues. **October–November:** Edison designs an improved phonoplex telegraph receiver and executes five patent applications covering phonoplex and multiple telegraph inventions. **November:** Ezra Gilliland joins Edison in New York City as a business and inventive partner. **8–10 December:** He designs an apparatus for experiments to discover an unknown force he calls XYZ. **13 December:** Edison leaves New York to visit the Miller family in Akron, Ohio.

1886   20 January: Edison purchases a home in Llewellyn Park, New Jersey. **30 January:** He merges the Electric Tube Company and the Edison Shafting Company into Edison Machine Works. **January:** Edison agrees to pay judgment and court costs in the long-running Seyfert lawsuit. **1 February:** He participates in public demonstration of the railway telegraph system on Staten Island. **16 February:** Edison incorporates Sims-Edison Electric Torpedo Company. **24 February:** He marries Mina Miller in Akron. **ca. 27 February:** Edison and Mina arrive in Florida on their honeymoon. **17 March:** He and Mina arrive at the Fort Myers estate, where they join daughter Marion and the Gillilands. **18 March:** Edison begins a prolific series of notebook entries witnessed by Mina. **March–April:** Edison plans the extensive planting and landscaping of the Fort Myers estate. **ca. 29 April:** He and Mina arrive in Akron. **ca. 4 May:** Edison and Mina move into "Glenmont," their Llewellyn Park home. **17 May:** Employees strike at Edison Machine Works in New York City. **11 June:** Edison begins lamp experiments at the Edison Lamp Company factory in East Newark and moves his laboratory there. **23 June:** He announces his decision to move Edison Machine Works to Schenectady. **2 July:** Edison incorporates the Edison United Manufacturing Company, which takes over the sales and installation work of the Edison Company for Isolated Lighting. **11–12 August:** Edison attends a meeting of the Association of Edison Illuminating Companies at Long Beach, Long Island. **5 October:** He commences work on a new "standard" phonograph. **25 October:** Edison drafts the first of several patent applications for the transmission and distribution of high-voltage currents. **ca. 10 November:** He drafts a comprehensive memorandum outlining the hazards and disadvantages of high-voltage alternating current. **26 November:** Edison Electric Light Company licenses Zipernowsky alternating-current transformer patents in the United States. **ca. 18 December:** Edison Machine Works moves to Schenectady. **23 December:** Edison Electric Light Company initiates an infringement suit against Westinghouse Electric. **30 December:** Edison becomes ill with pleurisy. **November–December:** He completes at least ten patent applications related to high-voltage electricity transmission and distribution.

1887   January: Edison is confined to the Llewellyn Park home with pleurisy. **9 February:** He leaves for Florida under the care of a nurse. **16 February:** He is appointed to a committee on uniform installation standards at the convention of the National Electric Light Association (Philadelphia). **1 March:** Edison drafts the first patent application on the pyromagnetic motor. **8–21 March:** He works on pyromagnetic motor experiments with Charles

Batchelor in Florida. **ca. 24 March:** Edison is treated surgically for facial abscess. **March:** Alfred Tate negotiates the license of Edison's stencil patents, allowing A. B. Dick Company to market the mimeograph machine. **6 April:** Edison instructs Batchelor to begin planning a new laboratory at West Orange, New Jersey. **ca. 8 April:** He suffers a recurrence of facial abscess. **30 April:** Edison returns from the South in improved health. **3 May:** Edison hires Henry Hudson Holly as the architect for the new laboratory. **7 May:** Edison begins sustained experiments on phonographs and recording cylinders. **19 May:** He approves architectural plans for a new laboratory, then leaves for Schenectady. **21 May:** Edison drafts a patent application for an improved phonograph cylinder and electrostatic playback mechanism. **5 June:** He has William K. L. Dickson begin sustained experiments on a magnetic ore separator. **ca. 5 July:** Edison shops for laboratory equipment in Philadelphia; he probably travels to Baltimore and Washington. **30 July:** Edison dismisses architect Henry Hudson Holly and replaces him with Joseph Taft. **July:** Edison begins planning the staff of the new laboratory and decides to add four small laboratory buildings. **15 August:** Edison's papers on the pyromagnetic dynamo and magnetic bridge are presented at a meeting of the AAAS. **August:** Edison solicits investment in a prospective company to manufacture inventions created at the new laboratory; he devises a five-wire electrical distribution system. **September:** He designates George Gouraud his agent for the phonograph in Great Britain. **1 October:** Edison contracts for the manufacture of phonographic dolls and toys. **8 October:** Edison incorporates the Edison Phonograph Company. **25 October:** He drafts a patent application on a battery device to convert alternating current to direct current for motors. **28 October:** He assigns all U.S. rights for the phonograph to the Edison Phonograph Company, which appoints Ezra Gilliland its sole general agent. **11 November:** Edison hosts visitors at laboratory from National Academy of Sciences meeting. **October–mid-November:** He gives newspaper interviews about the forthcoming phonograph and his new laboratory at West Orange. **15–25 November:** Edison makes extensive notes on experiments to be conducted at his new laboratory. **22 November:** He executes a patent application for the new phonograph. **19 December:** Edison replies to an official inquiry from New York State regarding the use of electricity for capital punishment. **23 December:** Edison lights Glenmont with electricity supplied by the laboratory.

1888 3 January: Edison drafts a long list of laboratory projects. **31 January:** He attends the opening of New York Electric Club. **January:** Edison discusses the formation of a partnership with Henry Villard to include research and manufacturing interests; he explores new ventures for financial support of the laboratory. He directs Jonas Aylsworth to begin a long series of experiments on the composition of phonograph cylinders. **ca. 4 February:** Edison agrees to a scientific and technical lecture series for laboratory employees. **ca. 7–13 February:** He travels to Chicago for a meeting of the Association of Edison Illuminating Companies. **ca. 27 February:** Photographic pioneer Eadweard Muybridge visits the laboratory. **January–February:** Edison renews the worldwide search for natural fibers to use as incandescent lamp filaments; he arranges funding from electrical manufacturing shops for experimental work. **February:** Edison investigates methods of duplicating phonograph recordings; the laboratory resumes development of the talking doll. **2 April:** Edison is elected a resident member of the New York Academy of Sciences. **April:** He directs research on the chemical treatment of tobacco leaves. **2 May:** Edison advises the Society for the Prevention of Cruelty to Animals on electrocution as a form of euthanasia. **3 May:** He incorporates Edison Phonograph Works. **11 May:** Edison opens his laboratory and exhibits the "improved" phonograph to representatives of the press. **12 May:** Edison exhibits the "improved" phonograph at the Electric Club in New York City. **14 May:** He postpones the dispatch of the latest phonograph machine to Great Britain; he authorizes contracts for the construction of a new phonograph factory. **31 May:** His daughter Madeleine Edison is born. **Late May–mid-June:** Edison redesigns the phonograph again. **ca. 13 June:** Samuel Clemens visits him at the laboratory. **16 June:** Edison records the first transatlantic phonographic message to George Gouraud. **ca. 17 June:** He dispatches the "perfected" phonograph to George Gouraud in London. **21 June:** Edison supervises the first animal electrocution experiment at his laboratory. **28 June:** Edison agrees in principle to sell his rights for the sale and distribution of phonographs in the United States and Canada to Jesse Lippincott and the prospective North American Phonograph Company. **June:** He considers forming the Amusement Phonograph Company for the commercial reproduction and distribution of recordings; he publishes "The Perfected Phonograph" in the *North American Review.* **14 July:** Jesse Lippincott organizes the North American Phonograph Company. **25 July:** Edison agrees with Walter Mallory to form the Edison Iron Concentrating Company for ore milling in Michigan, Wisconsin, and Minnesota. **26 July:** He designates John Birkinbine his consulting engineer for mining and ore milling. **July:** Edison adopts metallic "soap" as a new recording medium; he executes at least nine patent applications on sound recording. Animal electrocution experiments continue at the laboratory. Edison has the laboratory begin the production of a zinc solution for electric meters for central stations. The Edison Phonograph Works begins occupying new factory buildings. **9–25 August:** Edison suffers from dyspepsia and vacations at the Chautauqua Institution. **28 August:** Edison meets with Henry Villard about consolidating their electric lighting enterprises. **11 September:** He severs his friendship and business ties with Ezra Gilliland and John Tomlinson. **ca. 15 September:** Edison begins antimicrobial research related to outbreaks of yellow fever. **18 September:** He retains Sherburne Eaton as personal legal adviser in place of John Tomlinson. **26 September:** Edison sends Osgood Wiley to London to set up a demonstration ore separator. **September:** He continues trying to adapt phonograph cylinders for mailing. He quarrels with Edward Johnson over the phonograph business and the reorganization of the Edison

electric light and power interests. **8 October:** Edison executes a patent caveat for the kinetograph/kinetoscope, the first of four major caveats for early motion-picture machines. **31 October:** He places William Hammer in charge of his exhibit at the forthcoming Paris Universal Exposition. **ca. October–November:** Edison begins small-scale production of phonographs at Edison Phonograph Works in West Orange. **November:** He quarrels with longtime associate Sigmund Bergmann. He begins entertaining proposals to consolidate Edison electric light and power businesses. **5 December:** Edison attends animal electrocution demonstrations at a laboratory for the Medico-Legal Society committee investigating electrocution for capital punishment. **ca. 5 December:** Members of the American Institute of Mining Engineers visit the laboratory. **10–11 December:** Edison reaches a provisional agreement with Henry Villard for consolidating Edison electric light and power businesses. **ca. 22 December:** Edison leaves to spend the holiday season with the Miller family in Akron. **27 December:** The New Jersey and Pennsylvania Concentrating Works is organized.

1889 2 January: Edison General Electric Company is incorporated in New Jersey. **ca. 4 January:** Edison returns from Akron. **ca. 11 January:** Edison begins work on a new form of kinetograph/kinetoscope. **1 February:** Edison executes ten patent applications for phonograph improvements. **2 February:** He declines an invitation to meet George Westinghouse in Pittsburgh, Pennsylvania. **5 February:** Edison begins acquiring mining properties near Bechtelsville, Pennsylvania. **8 February:** He announces plans to disband the laboratory's chemical department. **20 February:** Edison hosts a session of the American Institute of Mining Engineers' annual meeting. **ca. 22 February:** He is injured in a laboratory accident. **February:** Edison considers (and rejects) patent-sharing arrangement with Thomson-Houston Electric Company. **6 March:** He dispatches William Hammer to the Universal Exposition. **ca. 9 March:** Marion Edison sails for France. **12 April:** Edison shuts down the Michigan experimental ore plant. **23 April:** The Edison General Electric Company is incorporated in New York State. **ca. 27 April:** Edison sends Albert Dick to Europe to investigate doll markets and manufacturing. **11 May:** Edison files suit against Ezra Gilliland and John Tomlinson over the phonograph contract. **22 May:** He attends the Westinghouse patent lawsuit in Pittsburgh. **May:** He begins constructing factories in Silver Lake, New Jersey. **ca. 11–15 June:** Edison inspects iron mining properties around Bechtelsville. **15 June:** He sends Adelbert Wangemann to Europe for phonograph exhibitions. **23 July:** Edison testifies as an electrical expert regarding the death sentence of William Kemmler. **July:** He and William K. L. Dickson begin using flexible film in kinetograph/kinetoscope experiments. **3 August:** Edison sails for France with Mina and Francis and Margaret Upton. **11 August:** Edison arrives in Paris. **13 August:** Edison and Mina ascend the Eiffel Tower for a reception in his honor and are photographed there by Count Giuseppe Primoli. **16 August:** He is made a Grand Officer of the Crown of Italy and attends a gathering at the Paris home of Roland Bonaparte. **19 August:** Edison meets with French president Sadi Carnot; he is received by the Académie des Science and later attends the banquet of the Société de Photographie. **22 August:** Edison visits Louis Pasteur at his Paris laboratory. **24 August:** He agrees to license the Lelande–Chaperon battery for the life of the patent; he and his party visit Versailles. **26 August:** Edison is honored by *Le Figaro* at a reception in Paris. **27 August:** Edison attends Buffalo Bill's Wild West Show near Paris. **10 September:** He is honored by Gustave Eiffel at the Eiffel Tower. **12 September:** Edison arrives in Berlin. **14–15 September:** Edison visits Werner von Seimens at his home in Charlottenburg. **17–18? September:** He attends the Congress of the Association of German Naturalists and Physicians in Heidelberg. **20–21 September:** Edison and Mina are in Brussels, Belgium. **21–26 September:** He and Mina visit London and stay at the country estate of Sir John Pender in Kent. **26 September:** He returns to Paris and is named a commander of the French Legion of Honor. **28**

September: Edison and Mina leave for New York. **6 October:** He arrives home and reviews changes to the kinetoscope and several experimental kinetograph films. **16 October:** He completes a new phonograph prototype. **October:** Edison begins to study the feasibility of a large hydroelectric plant at Niagara Falls. **ca. 4 November:** He dispatches a prospector to North Carolina to search for precious and base metals and minerals. **29 November–6 December:** Edison examines mine properties near Dover, New Jersey. **12–16 December:** He travels to western New York and Ontario with Alfred Tate. **Fall:** Edison begins marketing the Edison–Lalande battery. **23 December:** He testifies before a New York City grand jury about the safety of electric wires; he leaves with Mina for Akron. **27 December:** Edison receives a report from Sherburne Eaton on proposed agreements for the organization of the Edison United Phonograph Company. **December:** Edison purchases a large mineral collection for the laboratory. He begins to electrify a portion of the Orange Street railway line.

1890 ca. 1 January: Marion Edison contracts smallpox while traveling in Europe; her grave illness and yearlong convalescence abroad trigger a rift with her father and stepmother. **February:** The Automatic Phonograph Exhibition Company is organized to market the coin-in-the-slot phonograph. **30 April:** Edison closes the experimental ore milling plant at Bechtelsville. **March–May:** To reduce expenses, Edison progressively lays off about one-third of his laboratory staff. **3 August:** Edison's son Charles is born. **August:** Edison purchases property in Silver Lake (now the Bloomfield–Belleville area) for the Edison Manufacturing Company. **August-September:** Works on developing electric railway/streetcar motor and transmission systems, including spending several weeks at the former Edison Machine Works in Schenectady. **Summer:** He completes construction of the first of his iron concentration plants in Ogden (later Edison), New Jersey. **September:** Edison purchases the Ogden Iron Works. **October:** He reaches an agreement with the Edison General Electric

Company for support of his research on electric light and power.

1891 January: Edison travels to Port Huron for the funeral of his brother William Pitt Edison. **February:** He returns to the Edison General Electric Company plant in Schenectady for additional experiments on streetcar motors. **March–April:** He builds a test track at the West Orange laboratory for a low-voltage streetcar system. **28 April:** The New York Concentrating Works is incorporated. **12 May:** He travels to Chicago for discussions regarding the electric lighting business and plans for the 1893 Columbian Exposition. **20 May:** Edison demonstrates the kinetoscope at the West Orange laboratory for the Federation of Women's Clubs. **June–December:** He spends most of his time at the New Jersey and Pennsylvania Concentrating Works plant in Ogden. **14 July:** The primacy of Edison's lamp patents is upheld in the federal court decision of *Edison Electric Light Co. v. U.S. Electric Lighting Co.* **24 August:** Edison executes patent applications for the kinetoscope and kinetograph.

1892 March: Edison begins experimenting with the composition and production of iron ore briquettes at Ogden. **15 April:** Edison General Electric merges with Thomson-Houston Electric to form General Electric Company, ending Edison's decades of leadership in the electrical industry and marking a major transition in his career. **August–December:** Edison splits his time between Ogden and West Orange. **14 November:** Edison closes the Ogden plant for repairs and modifications, the first of several such shutdowns.

1893 Edison splits his time between Ogden and West Orange. **February:** Edison completes construction of the laboratory's Black Maria motion picture studio, which becomes fully operational in May. **9 May:** He displays his standard peephole kinetoscope (vertical-feed, 1½-inch film width) at the Brooklyn Institute of Arts and Sciences. **August–September:** He joins members of the Miller family on a visit to the Columbian Exposition in Chicago. **29**

December: Edison executes a patent application for the "Giant" ore crushing rolls.

1894 Edison spends most of his time during the year at the Ogden plant with frequent trips to West Orange. **January:** William K. L. Dickson produces *Edison Kinetographic Record of a Sneeze*, the first motion picture to receive a copyright. Dickson and Theodore Heise go on to copyright approximately seventy-five motion pictures in 1894. **14 April:** The first commercial viewing of the peephole kinetoscope is held by the Holland Brothers at 1155 Broadway, New York City. **April:** John Randolph succeeds Alfred Tate as Edison's private secretary. **Summer:** Edison reduces his laboratory staff by more than half. **21 August:** The North American Phonograph Company enters receivership. **December:** Edison closes the Ogden plant for repairs and design modifications.

1895 Edison spends most of his time during the year at the Ogden plant with frequent trips to West Orange. **Summer:** Edison experiments with the mass production of iron-ore briquettes suitable for shipping and use in blast furnaces (development continues through early 1897). **October:** He resumes work on squirted cellulose lamp filaments under a contract with General Electric Company. **1 October:** Edison's daughter Marion marries Karl Oscar Oeser in Dresden, Germany.

1896 Edison spends most of his time during the year at the Ogden plant with frequent trips to West Orange. **27 January:** Edison organizes the National Phonograph Company. **January–March:** He experiments with X-rays and sends a completed X-ray fluoroscope to Columbia University physicist Michael Pupin. **26 February:** Samuel Ogden Edison, the inventor's father, dies in Norwalk, Ohio; Edison attends the funeral. **March–July:** He closes the Ogden plant for modifications. **23 April:** The Edison Vitascope, a motion picture projector invented by Thomas Armat, debuts at Koster and Bial's Music Hall in New York City. **April:** Edison tests his gold-ore separation process on placer samples from the Ortiz

Mine in Dolores, New Mexico. **November:** He introduces the Edison Home Phonograph, an inexpensive phonograph with a spring-motor drive. **December:** Edison travels to the Crane Iron Works in Catasauqua, Pennsylvania, to observe test runs of pig iron produced from his iron-ore briquettes.

1897 Edison spends most of his time during the year at the Ogden plant with frequent trips to West Orange. **16 July:** Edison executes a patent application for the three-high crushing rolls in his ore milling process. **Summer:** The Ogden plant is again closed for repairs and modifications. **30 November:** Edison's own motion picture projector, the projectoscope or projecting kinetoscope, has its first commercial exhibition. **December:** Edison begins a series of lawsuits alleging patent infringement by competitors in the motion picture industry.

1898 Edison spends most of his time during the year at the Ogden plant with frequent trips to West Orange. **June:** Edison's son William Leslie joins the U.S. Army and will serve in the Spanish-American War. **10 July:** Edison's son Theodore Miller Edison is born and named after Mina Edison's brother, who died two days earlier in the Spanish-American War. **20 December:** Edison shuts down his ore milling plant at Ogden to repair the machinery and build more worker housing.

1899 **January:** Edison designs a long rotary kiln for making cement. **17 February:** Lewis Miller (Edison's father-in-law) dies; Edison attends his funeral in Akron. **19 February:** Thomas Alva Edison Jr. marries Marie Louise Toohey in the Roman Catholic Church, two days after announcing they had secretly married in November (the couple will separate within two years). **15 April:** Edison organizes the Edison Portland Cement Company. **18 April:** The Edison-Saunders Compressed Air Company is incorporated in New Jersey. **Summer:** He begins experimental work on storage batteries. **7 November:** William Leslie Edison marries Blanche Travers against his father's wishes.

1900 **31 January:** Edison's sister Marion Edison Page dies; Edison attends her funeral in Milan, Ohio. **March:** Edison vacations with his family in Florida, staying at the Tampa Bay hotel rather than their winter home in Fort Myers. **August:** Edison completes construction of a long rotary kiln at the Edison Portland Cement Company. **Summer:** Edison's mill for the concentration of gold ore begins operating at the Ortiz Mine in New Mexico. **September:** Edison stops operations at New Jersey and Pennsylvania Concentration Works in Ogden because of market conditions for iron. **1 November:** Edison shuts down the Ortiz Mine because of its ore's poor quality.

1901 **January–February:** Edison supervises construction of the Edison Portland Cement Company works at Stewartsville, New Jersey. **Mid-February:** He opens a motion picture studio at 41 East 21st Street in Manhattan. **27 February:** Edison begins family vacation in Fort Myers and visits his winter home there ("Seminole Lodge") for the first time since 1887. **14 May:** Edison receives threatening letters demanding $25,000 in gold, "or we will kidnap your child." He hires Pinkerton detectives, and the plot is stopped. **27 May:** He organizes the Edison Storage Battery Company. **25 July:** Edison begins a six-week tour of upstate New York and Canada, during which he visits Niagara Falls and the Pan-American Exposition in Buffalo with his family before traveling by ship across the Great Lakes to Sudbury, Ontario, with his wife Mina.

1902 **January:** Edison puts "moulded" records on the market. **3 March–2 April:** Edison and his family vacation in Fort Myers. **25 April:** The Dunderland Iron Ore Company Ltd is organized by the Edison Ore Milling Syndicate, Ltd, to exploit Edison's ore processing technology in Norway. **2 May:** The Mining Exploration Company of New Jersey is incorporated to find and develop nickel deposits for Edison's storage battery. **9 May:** Lord Kelvin (William Thomson) visits the West Orange laboratory. **May:** Edison conducts road tests of electric vehicles equipped with his storage batteries. **August:** Edison begins commercial production of cement at the mill in Stewartsville.

1903   January: Edison starts production of his "E"-type alkaline storage battery. **21 February–11 April:** Edison and his family vacation in Fort Myers. **2 March:** An explosion at the Edison Portland Cement Company plant kills eight workers. **8 June:** Edison signs an agreement with Thomas Edison Jr. whereby his son will not use his name in any business enterprise and will, in return, receive a weekly allowance of $35. **Fall:** There is labor unrest at the Edison works in Stewartsville and West Orange. **10 December:** Edison writes to President Theodore Roosevelt to influence the U.S. Patent Office in its judgment on the merits of storage battery patents by Edison and Ernest Waldemar Jungner of Sweden; the latter is eventually invalidated. **December:** The Edison Manufacturing Company releases its film *The Great Train Robbery*, directed by Edwin S. Porter; it becomes a big commercial success.

1904   10 February: Edison dismisses chemist Martin A. Rosanoff, effective March 2, declaring that he is "the last foreigner in the chemical line that I shall hire. I prefer bright young Americans." **29 February–10 April:** Edison and his family vacation in Fort Myers. **April:** The Edison Manufacturing Company films reenactments of the Russo-Japanese War for exhibition in the United States. **September:** Edison authorizes longtime associate Sigmund Bergmann to organize a corporation for the manufacture of storage batteries in Germany. **October:** Laboratory employee Clarence M. Dally dies as the result of radiation burns sustained during X-ray experiments. **November:** Edison suspends the manufacture of his alkaline storage battery to investigate leaks and the loss of electrical capacity.

1905   23 January: Edison has surgery to drain a mastoidal abscess. **7 March:** J. P. Morgan visits the laboratory to discuss Edison's improved storage battery and the formation of European companies for its manufacture. **Winter:** Edison forgoes his annual vacation in Florida because of work on the storage battery.

Summer: He begins a series of experiments using perforated tubes holding nickel flake as the positive electrode in his storage batteries; tests continue for a decade. **September:** Edison sends form letters to telegraph operators across the United States seeking information about deposits of cobalt ore for use in his storage battery.

1906   25 January: Edison wins a thirty-year lawsuit against Jay Gould's Atlantic & Pacific Telegraph Company for infringing his automatic telegraph patents but receives only $1 in damages. **17 February:** Marie Louise Toohey Edison dies; Edison pays for the funeral and the obituary in the *New York Herald*. **28 February–8 April:** Edison and his family vacation in Fort Myers. **March:** Edison's attorney Frank Dyer purchases on his behalf a farm in Burlington, New Jersey, for Thomas A. Edison Jr., who was recently released from a sanitarium for the treatment of alcoholism. **May:** Edison travels by car through North Carolina and other southern states prospecting for cobalt ore. **7 July:** Thomas A. Edison Jr. marries Beatrice Heyzer Montgomery, the two having been living under the assumed names of Burton and Beatrice Willard. **October:** Edison announces a plan to develop molds to make an entire house out of poured concrete.

1907   11 February: Edison declares his intention to give up commercial projects and devote himself to scientific laboratory research. **27 February–22 April:** Edison and his family vacation in Fort Myers. **5 March:** A federal appeals court affirms Edison's patent on a motion picture camera, increasing his control over American film production. **July:** Edison transfers the production of motion pictures from Manhattan to a new studio in the Bronx. **27 September:** Edison announces his decision to shut down the Darby Mine in Ontario because he no longer uses cobalt for his storage battery.

1908   8 January: Edison signs a patent cross-licensing agreement with the North American Portland Cement Company. **17 January:** Edison incorporates the Edison Business

Phonograph Company to exploit an improved dictating machine. **17 February:** Private secretary John F. Randolph dies by suicide; he is succeeded by Harry F. Miller. **23 February:** Edison enters the Manhattan Eye Ear & Throat Hospital; he has two additional operations on his left ear and remains there until 10 March. **ca. 19 March–28 April:** Edison vacations in Fort Myers with his wife and their children. **August:** Edison vacations with members of his family in the Pacific Northwest and Canadian Rockies. **1 September:** Edison takes control of the Lansden Company, which manufactures electric vehicles using Edison storage batteries. **9 September:** The Motion Picture Patents Company is incorporated by the Edison Manufacturing Company and the American Mutoscope & Biograph Company. **1 October:** Edison introduces Amberol cylinder records, which increase playing time from two to four minutes. **16 November:** Frank L. Dyer hires a personal attendant for Edison to "safeguard him from possible cranks and other people who might annoy him."

1909   February: Edison promises to pay Jonas W. Aylsworth $25,000 and Walter H. Miller $10,000 if they can develop a 400-thread cylinder record. He separately agrees to loan his son William Leslie Edison $150 to move to a house in the country. **18 February–mid-April:** Edison and Mina vacation in Fort Myers. **June:** Edison dictates personal reminiscences to Thomas C. Martin for an authorized biography. **1 July:** He begins commercial manufacture of the new "A"-type alkaline storage batteries. **December:** Edison begins to develop a disc record and phonograph.

1910   1 January: Edson's longtime friend and collaborator Charles Batchelor dies. **January:** Edison plans to establish an Engineering Department at the West Orange laboratory to centralize research and development for the numerous Edison companies. **February–April:** Edison and Mina vacation in Fort Myers with their children. **March:** A streetcar powered by Edison storage batteries begins operating on a crosstown line in New York City. **May:** Edison exhibits a scale model of his poured concrete house at New York's Madison Square Garden. **18 July:** He arranges with Miller Reese Hutchison to develop storage batteries for submarines. **26 August:** Edison demonstrates his kinetophone, or "speaking pictures," for the press at his laboratory. **17 September:** Two electric vehicles equipped with Edison storage batteries leave New York on a promotional tour, ending with an ascent of Mount Washington in New Hampshire. **October:** Edison receives national attention after making statements to the press revealing his unorthodox religious beliefs, including his skepticism about the existence of an immortal soul.

1911   24 January: Edison celebrates the silver anniversary of his marriage to Mina Miller Edison. **January:** He speaks about the immorality of the soul in the *Columbian Magazine*. **January–February:** He executes six successful applications relating to disc records. **28 February:** The National Phonograph Company is reorganized and incorporated as Thomas A. Edison Inc. **February–May:** Edison executes four successful patent applications for improvements in the reproducer for his disc phonograph. **23 March:** He hosts Count Lev Lvovich Tolstoy, son of the Russian novelist, at the West Orange laboratory. **10 July:** Edison shows his disc phonograph for the first time in public at the annual convention of the National Association of Talking Machine Jobbers, but sales do not begin for another year. **August–September:** Edison and Mina tour Europe with their children. **October:** He returns to his office and becomes actively involved in the selection of music and recording artists for his disc records. **October–November:** He places advertisements in *Iron Age* and other trade journals and newspapers soliciting investors and promoters for his "country house lighting system"—a plan to use storage batteries to illuminate rural homes located beyond gas and electric mains. **2 November:** He reviews the battleship fleet at the Brooklyn Navy Yard with President William Howard Taft. **11 November:** He makes Miller Reese Hutchison his personal representative at the West Orange laboratory. **27 November:** Edison calls on President Taft at the White House and visits the Washington

Navy Yard. **5 December:** He hosts about 200 officers and men from the Brooklyn Navy Yard who witness the first public demonstration of his new submarine battery. **30 December:** Edison executes a successful patent application for a small storage battery suitable for portable lamps that will subsequently be used in his miner's safety lamp.

1912   10 February: Edison celebrates his sixty-fifth birthday (11 February) at Glenmont with nearly forty acquaintances, many of them connected with his early work in electric lighting; he is presented with a loving cup and commemorative photographs. **March–April:** Edison, Mina, and their children vacation in Florida. **20 May:** Edison executes a successful patent application for an automobile starter motor and battery. **5 July:** He demonstrates his disc phonograph to approximately 100 jobbers at 10 Fifth Avenue in New York City; he also exhibits his home kinetoscope and improved business phonograph. **16 July:** He attends the first annual picnic and games day for his employees at Olympic Park in Irvington, New Jersey; the event will become known as Edison Field Day. **12 August:** He appoints Miller Reese Hutchison chief engineer of the West Orange laboratory, replacing Donald Bliss. **ca. 11 September:** Edison and six employees (nicknamed the "Insomnia Squad") begin a five-week effort to perfect the process for mass-producing disc records. **19 September:** He announces his support for Progressive Party presidential candidate Theodore Roosevelt and for women's suffrage. **7 October:** Edison begins shipping Blue Amberol cylinder machines and records to phonograph dealers. **22 October:** He attends the funeral of his mother-in-law, Mary Valinda Miller, in Akron. **25 October:** Edison executes three successful patent applications on processes for the mass production of disc records—the culmination of the work of the Insomnia Squad. **October:** The disc phonograph is exhibited at the Boston Electric Show in advance of marketing nationwide. **9 November:** Frank Dyer resigns as president of Thomas A. Edison Inc., allowing Edison to take control (which he holds until 1926). **29 November:** Edison borrows $500,000 from Henry Ford to construct a factory for storage batteries; Ford will eventually loan him $1.2 million for the project. **November:** Edison begins marketing the home projecting kinetoscope.

1913   2 January: Edison publishes his views on patent law in an article in *Leslie's Weekly*. **3 January:** He gives the first public demonstration of the kinetophone at the West Orange laboratory. **23 January:** The American Museum of Safety awards Edison its Rathenau Medal for developing a non-sparking, battery-powered safety lamp for mines and other enclosed spaces. **17 February:** Edison introduces talking pictures to American theatergoers by attending a performance of his kinetophone in New York City. **May:** He is named the "most useful" man in America by a survey of readers of *Independent* magazine. **August–September:** Edison takes an automobile trip through New England but falls ill and returns home early. **18 October:** With his wife Mina and son Charles, Edison leaves West Orange for a two-week vacation in Canada and the upper Midwest; they also visit Henry and Clara Ford in Detroit. **Fall:** Demonstrates educational motion pictures to an audience of distinguished teachers at the West Orange laboratory.

1914   23 February–17 April: Edison and his family vacation in Fort Myers and the Everglades with the Ford family and John Burroughs. **12 May:** Inspired by Henry Ford's anti-cigarette campaign, Edison bans cigarettes from all his plants (although he continues to smoke cigars and chew tobacco). **17 June:** His daughter Madeleine marries John Eyre Sloane, a Roman Catholic, against her parents' wishes. **8 September:** Responding to the outbreak of war in Europe and consequent embargoes, Edison announces plans to build a plant to manufacture phenol and other chemicals. He initiates a series of recitals and demonstrations of the Diamond Disc phonograph in New York, Boston, Philadelphia, and other cities. **9 December:** An explosion in the Film Inspection Building triggers a conflagration that destroys or damages more than half of the buildings in the West Orange laboratory complex. **10 December:** Edison hosts Henry Ford at Glenmont; he informs the press that the fire will cause "no interruption in my experiments." **December:** He suspends production of cement in his Stewartsville plant because of falling prices and low profits; it does not reopen until the spring of 1916.

1915 22 February: Edison opens his benzol plant at the works of the Cambria Steel Company in Johnstown, Pennsylvania; he claims that the factory was "finished in thirty working days after breaking ground." **1 March:** Edison announces his new divisional policy for Thomas A. Edison Inc. **19 March:** He begins construction of Phenol Plant No. 2 at Silver Lake; around the same time, he begins erection of a plant for making aniline from benzol. **10 April:** He opens his rebuilt phonograph factory eighteen weeks after half of the West Orange complex was destroyed by fire. **May:** Edison initiates a series of phonograph recitals at churches, hospitals, schools, police and fire departments, fraternal lodges, and other organizations throughout the country. **7 July:** Secretary of the Navy Josephus Daniels invites Edison to head the Naval Consulting Board; Edison accepts soon afterward. **9 August:** Edison begins a series of nationwide "tone tests," in which recording artists alternate live performance on a darkened stage with Edison Diamond Disc recordings and challenge audiences to detect the difference. **6 October:** He attends the first meeting of the Naval Consulting Board in the office of Josephus Daniels. **14 October:** With Mina, Edison departs for San Francisco, where he will be honored at the Panama-Pacific International Exposition. **October:** Edison meets Luther Burbank for the first time, visits the International Exposition with Henry Ford, and travels to Los Angeles and San Diego.

1916 15 January: An explosion aboard the U.S. Navy's E-2 submarine in the Brooklyn Navy Yard kills five men and injures ten others; the cause is attributed to hydrogen gas emitted by the recently installed Edison batteries. **4 March:** Edison's first grandchild, Thomas Edison ("Teddy") Sloane, is born. **15 March:**

Edison testifies before a congressional committee in support of an appropriation of $1.5 million for a naval research laboratory. **21 March–23 April:** Edison travels to Florida and vacations at his winter home in Fort Myers. **3 April:** The cement plant at Stewartsville reopens after being idle for sixteen months. **10 May:** Edison endorses Theodore Roosevelt for president but later throws his support to Woodrow Wilson after Roosevelt fails to win the Republican nomination. **13 May:** Edison and other members of the Naval Consulting Board march in a massive Citizens Preparedness Parade up Fifth Avenue in New York City. **31 May:** Henry Ford becomes Edison's neighbor in Fort Myers after completing negotiations with New York City businessman Robert W. Smith for the purchase of The Mangoes, the estate adjoining Edison's Seminole Lodge. **12 June:** Charles Edison is elected chairman of the board of directors of Thomas A. Edison Inc. **28 August:** Edison leaves West Orange for a camping trip in the Adirondack and Berkshire mountains with Harvey Firestone, John Burroughs, and Henry Ford (the "Vagabonds"). **3 September:** Edison announces, through the Democratic National Committee, his support for the reelection of President Woodrow Wilson. **31 October:** He travels by train to Detroit, where he visits Henry Ford. **9 December:** After the Naval Consulting Board recommends Annapolis as the site of a new naval research laboratory, Edison presents a minority report in favor of Sandy Hook, New Jersey. **29 December:** He falls ill at the laboratory after inhaling nitrous acid fumes and is confined to his house for almost two weeks.

1917 6 January: Consulting engineer Lamar Lyndon, hired by Thomas A. Edison Inc. to investigate the E-2 explosion, presents his final conclusion that the Edison battery "is the most suitable storage battery now obtainable for service on submarine vessels." **8 February:** Edison begins devoting most of his time to experiments for the U.S. government in a laboratory established in a casino on Eagle Rock Mountain in West Orange; over the course of two years, he will produce more than forty war-related inventions. **13 March:** He is elected "president for life" by the Naval Consulting Board to dispel rumors that he plans to resign from the board. **March:** He erects a small structure on the pier at the Sandy Hook Proving Grounds to conduct experiments for the U.S. Navy. **12 July:** Participates in tests for the U.S. Navy at Sandy Hook. **July:** Develops an ironclad waterproof army and navy phonograph for use by soldiers in the field. **20 August:** Meets with President Woodrow Wilson and Secretary of the Navy Josephus Daniels at the White House. **21 August–6 October:** Edison spends six weeks conducting experiments aboard the USS *Sachem* in Long Island Sound. **9 October:** Departs for Washington, D.C., and sets up an office in the Navy Annex in a room once occupied by Admiral George Dewey. Research operations on Long Island Sound continue until early December. **Mid-December:** Returns to West Orange and spends the holidays with his family.

1918 2 January: Returns to his office at the Naval Annex in Washington, D.C. Charles Edison becomes vice president and general manager of Thomas A. Edison Inc. **24 January:** The Edison Pioneers, an organization of former associates in the early electric light and power industry, is founded. **28 January:** Edison and Mina depart Washington for Florida—he to the Naval Station at Key West for research until late April and she to their home in Fort Myers. **27 March:** Charles Edison marries Carolyn Hawkins at Seminole Lodge in Fort Myers. **30 March:** Edison ends his involvement in the motion picture business by selling his Bronx studio to the Lincoln & Parker Film Company (though he will reacquire his assets after the company's bankruptcy in 1919). **1 April:** Joins his son Theodore in Man Key, an uninhabited island in the Florida Keys, where they conduct tests for the U.S. Navy. **21 April:** Edison's second grandchild, John Edison ("Jack") Sloane, is born. **4 May:** He visits the Naval Experimental Station in New London, Connecticut, with Secretary of the Navy Josephus Daniels. **9 May:** Edison visits Washington, D.C. **1 June:** Travels by automobile to Pittston, Pennsylvania, to visit the Black Diamond Powder Company—manufacturers of

high-grade explosives. **9 June:** Visits Washington, D.C., with his son Theodore to confer with government officials about the U-boat problem. **16 August–2 September:** Edison begins a two-week camping trip in the Shenandoah Valley and Great Smoky Mountains with John Burroughs, Harvey Firestone, and Henry Ford. **15 October:** Edison is confined to bed with a cold; family members deny that the malady has any resemblance to the Spanish influenza sweeping the nation.

1919 10 February: Departs from West Orange with Mina, their daughter Madeleine, and her husband and two children for a winter stay at Fort Myers. On the way to the train station, he meets for half an hour with the Edison Pioneers at their second annual Edison birthday luncheon in Newark. **11 April:** On his way back to New Jersey, he stops briefly in Savannah, Georgia, where he gives a press interview and comes out in support of Woodrow Wilson's plan for a League of Nations. **18 April:** Edison initiates a search in the United States and United Kingdom for "old songs" with "fine melodies" to be made into recordings; by the summer, he spends more than two hours per day reviewing the songs and finding "about one beautiful tune in 300." **21 June:** Attends the eighth annual Edison Field Day events, where he gives a press interview about his support for the League of Nations. **3 August:** Edison leaves West Orange for several weeks of camping in the mountains of Vermont and New Hampshire with Burroughs, Ford, and Firestone—the Vagabonds. **15 September:** Edison begins a campaign to produce a starter battery for Ford automobiles. **1 October:** He sells his studio in the Bronx, along with rights to his films, to motion picture producer Robert L. Giffen for $75,000. **13 November:** Edison sends the U.S. Navy his bill for experimental work, thus ending his involvement in military research.

1920 February–April: Edison and Mina are in Fort Myers and are visited by some of her extended family. **October:** Confronted with a postwar economic downturn, Edison dramatically thins the managerial and manufacturing workforce of Thomas A. Edison Inc. over the strong objections of his son Charles.

1921   25 January: Edison resigns from the Naval Consulting Board after a long dispute over the mission and location of the proposed naval research laboratory. **11 May:** Questions from Edison's test for college graduates seeking employment at his company are published by the *New York Times*, leading to extensive criticism and public debate about them. **July:** Edison goes camping in Maryland with Harvey Firestone and President Warren G. Harding. **November:** He travels by train with Henry Ford to Muscle Shoals, Alabama, making several stops en route.

1922   March–April: Edison visits Fort Myers with Mina, Charles Edison and his wife Madeleine Edison Sloane, and extended family members. **May:** Edison designs a plan for reforming the monetary system by extending credit to farmers based on the cash value of their crops. **June:** He receives an honorary doctor of science degree from Rutgers University.

1923   March–May: Edison and Mina visit Fort Myers with two grandchildren; they also receive Harvey Firestone and Henry Ford. **1 August:** Edison attends the Ohio funeral of President Warren G. Harding. **August:** After the funeral, Edison and Mina vacation in Canada and visit the Ford family in Detroit.

1924   February–April: Edison and Mina are in Fort Myers, where they receive extended family members and Firestone. **28 August:** Edison consolidates the Edison Phonograph Works into Thomas A. Edison Inc.

1925   February–April: Edison and Mina go to Fort Myers with their son Theodore and are visited there by Ford and Firestone.

1926   1 February: Edison signs his will, leaving the bulk of estate (mainly corporate holdings) to his sons Charles and Theodore. The same day, he sells all his domestic and foreign patents to Thomas A. Edison Inc. for $78,200

in cash. **February–April:** Edison and Mina are in Fort Myers with Madeine and her sons, and they again host extended family, Firestone, and the Fords. **2 August:** Edison steps down as president of Thomas A. Edison Inc. (his son Charles taking his place) and becomes chairman of the board. **October:** Edison introduces long-playing disc records in an attempt to save his declining phonograph business. He also begins offering attachments so that his phonographs can play competitors' laterally cut records.

1927   February–May: Edison and Mina are in Fort Myers and again host members of their extended family. **April:** The National Academy of Sciences elects Edison as a member (after two failed attempts in prior years). **July:** Edison organizes the Edison Botanic Research Corp. (with Henry Ford and Harvey Firestone) to develop a process for making rubber from plants native to the United States.

1928   Edison spends much of his time searching for and experimenting with plants from which to develop domestic sources of rubber. **January–June:** Edison and Mina are in Fort Myers and again spend time with Firestone and the Fords. **April:** Edison reluctantly agrees to Charles Edison's desire for Thomas A. Edison Inc. to enter the radio business. **20 October:** Edison receives a special Congressional Medal from Treasury Secretary Andrew W. Mellon in a ceremony broadcast nationwide on the radio.

1929   Edison spends much of his time searching for and experimenting with plants from which to develop domestic sources of rubber. **January–June:** Edison and Mina are in Fort Myers with Charles and his wife Carolyn. **February:** Edison celebrates his eighty-second birthday at Fort Myers with Ford, Firestone, and President-elect Herbert Hoover. **25 July:** He attends the Lewis Miller centenary celebration at the Chautauqua Institution with his family and Henry Ford. **21 October:** Edison attends the opening and dedication of Greenfield Village and the Henry Ford Museum as part of ceremonies marking the fiftieth anniversary of his incandescent light; President Hoover and more than 500 invited dignitaries attend. **28 October:** Edison stops production of phonograph records, beginning a shift to making radios. **5 December:** Edison and Mina return to Fort Myers.

1930 Edison spends much of his time searching for and experimenting with plants from which to develop domestic sources of rubber. **January–June:** Edison and Mina stay in Fort Myers and are visited by their by son Charles and his wife, daughter Marion, and various extended family members.

1931 **6 January:** Edison executes final patent applications. **January–June:** Edison and Mina are in Fort Myers with family members. **11 June:** He addresses an electric lighting convention in Atlantic City, New Jersey, by radio broadcast, offering Americans encouragement in the face of the Great Depression; the remarks are his last in public. **15 June:** With Edison in poor health, he and Mina begin the homeward journey. **1 August:** Edison collapses at Glenmont. **18 October:** He dies at home. **21 October:** Edison is buried in West Orange. At President Hoover's request, the nation turns off electric lights to observe one minute of darkness at 10:00 p.m. Eastern Standard Time.

Introduction

Most people think of Thomas A. Edison as the last great lone inventor. He is commonly identified as the inventor of the lightbulb, phonograph, and motion pictures. He is also associated with a host of other lesser-known inventions, such as the stock ticker. While Edison was not alone in contributing to these technologies, he did a great deal to create the electrical, sound recording, and motion picture industries. He also made significant innovations in telecommunications, battery technology, office machinery, the manufacture of Portland cement, and processes for working low-grade ores. He was able to contribute to such a wide array of industries specifically because he was not a lone inventor. At his workshops and laboratories in Newark, Menlo Park, and West Orange in New Jersey, Edison brought together teams of skilled research assistants and machinists. These teams allowed him to do more than any one person could do. In the process, he transformed invention by making it part of a larger process of research, development, and commercialization that we now call innovation. That transformation, as much as any single invention, has become a crucial feature of the modern world.

Edison's long career reflects the vast changes that took place in American society between the Civil War and the Great Depression. During these seven decades—the span of one long life—Americans increasingly moved from farms and towns to large cities, manufacturing grew from small shops into large factories, and business enterprises changed from small family firms to large stockholding corporations. These firms operated on a national scale and relied on telegraph and rail lines that crossed the continent and created coast-to-coast communications and distribution networks. Science and technology became more closely entwined as new electrical and chemical industries sprang up during what has been called a second industrial revolution. While Edison's work was central to these transformations, his success was made possible by the technological, economic, and social networks in which he was enmeshed. Edison's singular life and the web of connections around it help illuminate the United States of his time—and our own.

EDISON'S BOYHOOD

The foundation of Edison's success can be found in his youth and early career. Although the towns of his childhood—Milan, Ohio (population 1,500), where he was born on 11 February 1847, and Port Huron, Michigan (3,000), where the family moved in 1854—were small, they were local centers of commerce and industry, and Edison absorbed their culture of artisans and workshops. His father, Samuel, was a jack-of-all trades exiled from his native Canada due to his participation in an aborted revolution. His mother, Nancy, a former schoolteacher, provided his primary education. Edison briefly attended a private school in Port Huron but was otherwise taught by his mother until he reached high school age. His interest in science may have led to his brief attendance at the Union School, where he encountered

Richard Parker's *A School Compendium of Natural and Experimental Philosophy*, which included sections on electricity and the telegraph. Edison's education was also influenced by his father's library, which included many significant Enlightenment writers, including Thomas Paine, whose ideas about religion particularly impressed the young Edison. He also learned from the entrepreneurial ways of his father, whose many careers included shingle making, lumbering, running a grocery, and land speculation.

At the age of twelve, Edison worked briefly on his father's truck farm before obtaining a job selling candy, newspapers, magazines, and dime novels on the Grand Trunk Railway. While riding the train between Port Huron and Detroit, he printed his own newspaper, the *Weekly Herald*, in the baggage car, where for a time he also had a chemistry laboratory. To aid his chemical studies, he acquired an English translation of a textbook by German chemist Carl Fresenius. It was during this time that he first noticed the hearing loss that plagued him throughout his life. He later claimed that poor hearing was an advantage because it reduced distractions and enabled him to concentrate.

THE MAKING OF AN INVENTOR

Edison grew up with the telegraph, the first of the electrical industries that would transform the world. About the time he read Parker, he and a friend built a half-mile telegraph line between their houses. While working on the Grand Trunk, Edison began visiting with railway telegraphers along the line who taught him telegraphy. He soon was proficient enough to get a job in the local office, and between 1863 and 1867, he traveled the Midwest as an itinerant telegrapher. During these years, he became a very fast receiver, notable for his distinctive handwriting, and joined the ranks of elite press-wire operators who handled news dispatches. He also began experimenting with telegraph instruments and reading technical and scientific literature ranging from telegraph trade periodicals to Michael Faraday's *Experimental Researches in Electricity*. By the time Edison began working in Western Union's Bos-

ton office in 1868, he was thoroughly familiar with the science and art of telegraphy and had begun to learn the craft of invention.

In Boston, Edison encountered leading inventors, major manufacturing shops with skilled experimental mechanics, important industry officials, and capitalists looking for promising inventors and inventions. Here he found financial backers and mechanics able to help him realize his ideas. He acquired working space in the shop of Charles Williams, a leading telegraph manufacturer who also provided laboratory facilities to the prominent electrical inventor Moses Farmer. Edison soon applied successfully for two patents—a vote recorder for legislators and a printing telegraph for use in a stock quotation service. In early 1869, Edison resigned his operator's post to devote his time to invention.

One of his early inventions was a double transmitter for sending two simultaneous messages on a single wire. Edison traveled to New York City in early 1869 to experiment on this system with Franklin Pope, a prominent telegraph engineer with whom he formed a brief partnership. In New York, Edison found the movers and shakers of the telegraph industry. Western Union was headquartered in Manhattan, and the lawyers, bankers, and inventors at the heart of the business were collected there. The many creative telegraph inventors located in New York contributed to the development of long-distance technology and helped to create a distinctive urban telegraph industry serving the city's financiers and other businessmen. Edison's work on printing telegraphs for distributing financial information drew the attention of the Gold and Stock Telegraph Company, which sought to control the burgeoning market for quotation services at stock, gold, and commodity exchanges in New York and other major cities.

In February 1870, Gold and Stock acquired Pope and Edison's printing telegraph patents, and two of the company's directors contracted with Edison for an improved printing telegraph and a facsimile system. With money from these contracts, Edison opened his first telegraph manufacturing shop in Newark, New Jersey. That started a lifelong pattern in which he put his profits into resources to further his inventive processes. In whatever the mix of motives that drove his work—the joys of creativity and problem-solving, camaraderie, ego, or helping friends and humanity in general—he used money as a means rather than its own end. From the time he opened his Newark shop, he was never without a well-equipped shop and the services of highly skilled mechanics for rapid prototyping and development of his inventions.

In October 1870, he established a second and even larger shop, the American Telegraph Works, funded by the president of the Automatic Telegraph Company (for which Edison was developing a high-speed system). The following May, he became the "consulting Electrician and Mechanician" for Gold and Stock (*TAEB* 1:154). That move incidentally brought him within the purview of Western Union, which at the time was taking over Gold and Stock as part of its effort to control the distribution of commercial news.

Under an oral arrangement with Western Union president William Orton, Edison worked on multiple telegraph systems in 1873 and 1874, culminating in his quadruplex (four-message) telegraph. At the same time, he was improving his system for the rival Automatic Telegraph. As 1875 opened, he became embroiled in financier Jay Gould's attempt to build a competing network to Western Union's by acquiring rights to the automatic and quadruplex telegraphs, leading to years of litigation. Although Gould employed Edison for several months as an electrician for his Atlantic and Pacific Telegraph Company, the inventor soon returned to the Western Union fold by agreeing to a contract that assigned all his work in multiple telegraphy to the company. In retrospect, one can see in these early experiences and relationships in the high-tech telegraph industry—as well as his abundant curiosity and native creativity— the makings of success in the novel field of professional inventing.

CREATING A RESEARCH LABORATORY

As Edison shifted his focus from electromechanical printing telegraphs to high-speed automatic and multiple transmission systems, he began to transform his approach to invention. American telegraph instruments, including the Morse key-and-sounder system and printing telegraphs for financial reporting and other local telegraph systems, produced little in the way of complex electrical phenomena. British telegraphers, by contrast, confronted puzzling effects created by having to run their lines underground through cities (U.S. lines were not forced underground until the 1880s). And for two decades, British engineers had been forced to study electrical induction in the far-flung undersea cables through which Britain administered its empire. Edison had some knowledge of induction effects, but a trip to England in early 1873 gave him firsthand experience and brought him into contact with the sophisticated electrical community there.

The trip helped Edison understand how much he did not know about the electrical and chemical phenomena in cable and automatic telegraphy. Soon after his return home in mid-1873, he established his first real laboratory in a corner of his Newark telegraph works. There he undertook basic applied research related to his inventive work and cultivated an appreciation of fine test instruments and precise measurements. He soon boasted that his laboratory contained "every conceivable variety of Electrical Apparatus, and any quantity of Chemicals for experimentation" (*TAEB* 2:104). He was joined there by Charles Batchelor, who had been foreman of Edison's Newark shop. Batchelor had no experience with electrical technology when he started with Edison but had since gained a good deal of practical knowledge, and he possessed great skill for delicate and precise operations. He would remain Edison's principal assistant for the next twenty years.

In his new laboratory, Edison began to focus his experiments on electrical and electrochemical phenomena rather than on the electromechanical devices that had made his early reputation. His research broached fundamental scientific questions as he explored little-understood effects and proposed general theories to account for them. Despite this breadth of approach, however, he remained intent on producing new technology. His experiments were made in pursuit of a deeper understanding of the electrical and chemical action of telegraph devices, and he aimed always to apply his knowledge to developing new inventions.

By May 1875, Edison's success as a telegraph inventor gave him sufficient resources to turn to invention full-time. He expanded his laboratory to two floors of the Newark shop and handed all manufacturing activities to his partner Joseph Murray. By making his experimental machine shop entirely independent of manufacturing and incorporating it into a research laboratory, Edison completed the transition he had started after his return from England nearly two years prior. With the skilled workmen and tools from his Newark telegraph shops adapted solely to inventive work, he could rapidly construct, test, and alter experimental devices, significantly increasing the rate at which he could develop new inventions. In this new laboratory, Edison developed his first significant non-telegraph invention—the electric pen and press for creating multiple copies of handwritten documents. His experiments on acoustic telegraphy (a system for sending several messages of different frequencies over a telegraph wire) led to his controversial claim to have discovered a mysterious new form of energy in the ether. Only later was Edison's "etheric force" recognized as a form of electromagnetic radiation akin to radio waves.

THE MENLO PARK LABORATORY

Edison further expanded his capabilities in 1876 when he built his now-famous laboratory in Menlo Park, New Jersey. Notably, he left the financial and manufacturing centers of New York and Newark to do so, though he maintained close ties to the metropolis. The 25- by 100-foot wooden laboratory building housed a precision machine shop on the ground floor and a chemical and electrical laboratory on the second floor. Edison spent about $40,000 (more than $1 million today) for the machinery and scientific apparatus. Three machinists and two experimenters from Newark accompanied him to Menlo Park. No other inventor, either inside a great corporation like Western Union or as a freelancer, had assembled such a combination of material, intellectual, and craft resources. With it, the twenty-nine-year-old Edison aimed to do something equally novel: create a steady stream of original products by turning invention itself into an industrial process.

Edison's great success in the next five years would make his Menlo Park laboratory a model for others. The laboratory's influence was evident in the number of American and European scientists who visited and found Edison better equipped than they were. Inventors such as Alexander Graham Bell were spurred by what Bell called Edison's "celebrated laboratory at Menlo Park" (Bruce 1973, 355) to set up their own facilities. The Bell Telephone Company drew on Edison's example when it established an experimental shop headed by Edison's longtime friend and associate Ezra Gilliland.

The first year of work at Menlo Park focused on various systems of multiple telegraphy for Western Union. In January 1877, Edison proposed that the company support his machine shop with a weekly stipend, and he and President William Orton soon signed an agreement giving Western Union the rights to all of Edison's telegraph inventions in return for $100 a week in laboratory expenses. At Western Union's request, Edison and his staff turned their attention to the telephone—the speaking telegraph—which Alexander Graham Bell had unveiled in 1876. Bell's invention, though it transmitted the voice clearly, was too weak to use practically in the electrically noisy urban environment, nor could it send a signal any distance. Edison took a different approach to the problem of capturing the voice; "Bell," he said, "got ahead of me by striking a principle of easy application whereas I have been plodding along in the correct principle but harder of application" (*TAEB* 3:594). As the multiple telegraph research tailed off, Edison's telephone research intensified, culminating in the spring of 1878 in one of his most enduring products: the carbon-button transmitter, which became the industry standard.

In the process of his telephone research, Edison conceived of a device to record and repeat telephone messages. It became a separate invention for recording and reproducing sound that Edison called the phonograph. In early December 1877, he demonstrated his new talking machine at the offices of *Scientific American* in New York City. Although Edison found backers associated with Bell Telephone to form a company to exploit this new marvel, he was unable to transform this early exhibition machine into a commercial product. Nonetheless, the tinfoil-cylinder phonograph transformed him almost overnight into an international celebrity—the "Wizard of Menlo Park." Edison's familiarity with the press and his rapport with journalists enabled him to maintain and manipulate that celebrity for the rest of his life.

Edison's efforts on the telephone and phonograph began to alter the scale of work at Menlo Park. With funding from Western Union and the Edison Speaking Phonograph Company, he increased his staff from the original small group who had accompanied him from Newark to around twenty men by the spring of 1878. They included four experimenters, a couple of general laboratory assistants, six machinists, a patternmaker, a general handyman, a watchman, a bookkeeper, and a private secretary. Over the next two years, as Edison turned to research on electric lighting, Menlo Park would become a true research and development laboratory.

Financial support for work on electric lighting came from investors connected with Western Union (including partners in the investment banking firm of Drexel, Morgan & Company) who organized the Edison Electric Light Company. Between October 1878 and March 1881, Edison Electric provided $130,000 (about $4 million today) for research and development. Their funds enabled Edison to expand his laboratory facilities, enlarge his staff, and experiment simultaneously on each aspect of his new system.

Beginning in the fall of 1878, Edison added several machinists and college-educated experimenters. Among the latter were Francis Upton, who received the first master of science

degree from Princeton and did postgraduate work with Hermann von Helmholtz in Berlin, and two chemists with German doctorates. Other new hires included a scientific-instrument glassblower, a steam engineer, a draftsman, and general laboratory workers. As he shifted from research to commercial development of the electric light system in 1880, Edison expanded his staff to about sixty men by adding even more machinists, trained engineers, and experimenters. Unable to afford assistants idling as they awaited his directions, he learned to subdivide the work by assigning each detail of the system to a particular staff member or a team of researchers and machinists.

CREATING THE ELECTRICAL INDUSTRY

Spurred by recent developments in arc lighting for streets and public spaces, Edison began to investigate incandescent lighting for indoor spaces in the late summer of 1878. That September, he announced that he had solved the problem of lighting by incandescence, which had stymied years of efforts by other inventors. Unlike many scientific and technical contemporaries, Edison quickly realized that Ohm's and Joule's laws required a system of incandescent lighting to use high-resistance lamps in order to reduce the size and thus cost of copper conductors. This theoretical framework provided the goal, but the unexpected difficulty of reducing it to practice made Edison's September declaration wildly premature.

For nearly a year, Edison tried filaments of metal (primarily platinum) with a high melting point. To maintain the metal burners at the right temperature, he designed regulators to shut the current off before they reached their melting point. Discovering that his filaments were melting prematurely, he undertook a series of experiments to study how they reacted at the high temperatures needed for incandescence. This led him to observe tiny amounts of gas trapped in the metals, which he sought to remove by heating the filaments in a strong vacuum. He presented his work on occluded gases at the annual meeting of the American Association for the Advancement of Science in August 1879. By October, Edison and his staff had improved his vacuum technology sufficiently that he began to experiment with carbonized substances that would burn up if the vacuum were insufficient. Besides being much cheaper than platinum, carbonized materials had the high resistance needed to make his system economically viable. Over the next year, Edison and his staff sought to find carbonous materials best suited for use in his lamps, eventually settling on bamboo because of its long, contiguous fibers.

Besides the lamp, Edison had to design all the other elements of the system, which he modeled after the gas lighting systems used in large cities. Components included underground conductors, meters, and fixtures. He also required something unique to electrical systems: an efficient generator. In the fall of 1878, after experiments indicated that existing arc light generators were inefficient for incandescent lighting, Edison and his chief assistants (Charles Batchelor and Francis Upton) investigated contemporary designs and the principles of electromagnetism. Early the next year, Edison designed a dynamo that differed in important ways from commonly used electric generators.

Having developed the basic forms of his lamp and generator, Edison conducted a series of public demonstrations at Menlo Park at the end of 1879. Over the course of the next year, he and his staff developed each part of the system. They built a scale model to demonstrate to investors and government officials. They also turned a nearby building into the first lamp factory, with the laboratory supplying all the machinery and vacuum pumps. They began manufacturing bulbs in October 1880. By the end of that year, Edison was satisfied that he had a practical system and made plans to begin its commercial introduction by installing the first permanent central station on Pearl Street in Lower Manhattan.

COMMERCIALIZING ELECTRIC LIGHTING

It was one thing to design a system of interior lighting, but building and installing it in the streets and buildings of New York City required different skills, a constellation of new organizations, and still more money. The Pearl Street station was funded by Edison Electric investors (especially the Drexel, Morgan partners), who organized the Edison Illuminating Company of New York as a utility and invested in the Electric Tube Works, which manufactured and installed the underground conductors for the Pearl Street district. However, Edison was unable to convince these financiers to underwrite the other enterprises needed to manufacture the components of his system. Instead, he drew on his own resources and those of some of his closest associates (including Charles Batchelor, John Kruesi, Francis Upton, Edward Johnson, and Sigmund Bergmann) and directed Samuel Insull, his new private secretary, to oversee the financial arrangements. To get the new companies off the ground, Edison cashed out stock in companies formed to market his telephone and electric light in European countries, and he used his stock in Edison Electric and the Edison Illuminating Company of New York as collateral for loans from Drexel, Morgan. His partners also risked their own money, including shares of Edison Electric that Edison had given them. The manufacturing firms would come to include the Edison Lamp Company (with its works in Menlo Park), the Edison Machine Works, and Bergmann & Company (both in New York). Edison's outside investors did help to create the Edison Company for Isolated Lighting, which sold individual lighting plants. By May 1883, the Isolated Company had installed 330 plants in a wide range of businesses, including textile mills and other factories, hotels, restaurants, stores, theaters, steamships, and railway stations.

The manufacturing shops and the installation of isolated plants were crucial for the ongoing research and development of the electric lighting system. The Isolated Company collected reports on the operation of the system from its agents and customers in the United States and abroad that helped Edison identify technical problems. The company also created demand for the output of his factories, giving them work and experience as the larger task of wiring New York's financial district moved slowly ahead. Development work took place largely at the manufacturing shops. At the Machine Works, staff tested and improved dynamos and meters and tested equipment developed at the other shops. These items included conductors and junction boxes from the Electric Tube Works and accessories such as regulators, safety fuses, and lamp fixtures from Bergmann & Company. In September 1882, Edison closed his operations at Menlo Park. He set up a laboratory at the Bergmann shop in New York and moved lamp production to a larger plant in Harrison, New Jersey, just across the Hudson River.

Just as the lamp had been the most difficult feature of the system to invent, its manufacture required constant experimental efforts to improve materials, equipment, and processes. For that reason, Edison periodically found it necessary to spend weeks at a time at the lamp factory troubleshooting or experimenting. Unlike other parts of the system, advancements in lamp technology relied to a significant extent on improvements in manufacturing techniques. By the spring of 1884, more than 2,770 lamp experiments had been conducted at a cost of more than $70,000 (more than $2.8 million today). In June 1886, in the face of growing competition, Edison moved his laboratory from the Bergmann & Company shop to the lamp factory in Harrison.

While the isolated lighting business boomed, the central station business grew much more slowly. In part, this was due to the enormous cost of the Pearl Street station. Before it began operating in September 1882, it required more than a year and $300,000 (about $9 million today) to construct and equip the station, lay underground conductors, and wire buildings within its one-square-mile district. It could not easily be replicated in other cities. Because of the expense, Edison developed a cheaper distribution system for lower-density cities and towns, and he created an independent business, the Thomas A. Edison Construction Department, to market and install it. The new system included a more efficient three-wire distribution network that extended the distance over which a station could deliver energy, and dynamos that were smaller, cheaper, and more efficient than the

"Jumbos" at Pearl Street. He also accommodated small-town stations by developing a less expensive overhead distribution system used for all but four of the seventeen stations built by the Edison Construction Department.

Having proved the economic viability of the central station business, Edison negotiated an arrangement in which the Isolated Company took over the central station business, with Edison Electric receiving the right to purchase stock in the manufacturing shops and any patents obtained by the shops. These agreements were part of a larger reorganization of the Edison lighting companies that culminated with a successful proxy fight waged by Edison at the October 1884 meeting of Edison Electric that left his close associate Edward Johnson in charge of the Edison lighting business.

FAMILY CHANGES

The reorganization of the lighting business allowed Edison time to properly grieve the death of his wife Mary, who died unexpectedly at their Menlo Park home in August 1884. Mary's mother took charge of the children and helped keep house when the family moved back to New York in September. Edison's daughter Marion became his almost constant companion over the next several months. She accompanied him to Philadelphia for the International Electrical Exhibition and joined him in his New York laboratory at Bergmann & Company.

In late 1884, Edison's experimental work shifted from electric lighting to telephony. At the electrical exhibition in Philadelphia, he had met up with his old friend Ezra Gilliland, who was then in charge of American Bell Telephone Company's experimental shop in Boston. Gilliland encouraged Edison to turn his attention to two key problems facing his company: long-distance transmission and selective signaling (ringing only the call bell of the intended recipient). These problems became the focus of Edison's experimental work for the next several months, although he never reached a formal arrangement with American Bell.

In early 1885, Edison traveled with his daughter and Ezra Gilliland and Gilliland's wife

Lillian to cities in the Midwest where the two men had worked as telegraphers. They then took the train to New Orleans and attended the World's Industrial and Cotton Centennial Exposition. From there, they journeyed to Florida. Instead of visiting the eastern part of the state, where he and Mary had vacationed, Edison went down the Gulf Coast to hunt and fish with Ezra. Their final stop was in the small town of Fort Myers, where Edison and Gilliland decided to purchase land on which to build adjacent winter homes.

According to Marion, Edison asked Lillian Gilliland to find him a wife who could be a mother to his three children. At the end of June, the Gillilands invited Edison and several suitable young women to their summer cottage near Boston in Winthrop, Massachusetts. Among the guests was Mina Miller, daughter of Lewis Miller, an Ohio farm-equipment manufacturer and cofounder of the Methodist summer educational camp known as the Chautauqua Institution. Although Mina stayed only a short time, Edison could not get her out of his mind. He declared his attraction in a journal he kept as part of a parlor game, the idea being for guests to share their entries with the whole party. Soon after, Edison visited the Miller family at Chautauqua, New York. Mina then accompanied him, Marion, and the Gillilands to Niagara Falls, Montreal, and the White Mountains of New Hampshire, where Edison tapped out a marriage proposal in Morse code on the palm of her hand.

After returning to New York, Edison was joined in his laboratory by Gilliland, who left the Bell Company. They resumed work on a railway telegraph system and Edison's phonoplex (for sending multiple signals between intermediate stations over a single wire). Edison also began what would become a long-term experimental program to discover an unknown form of energy akin to the etheric force, which he named "XYZ."

Edison and Mina married on 24 February 1886 at her parents' home in Akron, Ohio. They then traveled with Marion to Fort Myers, where the Gillilands joined them. Although the small laboratory next to his new winter home was not yet ready, Edison filled six notebooks with entries about ideas for experiments, all witnessed by Mina. Most of Edison's notes focused on electric lighting, telegraphy, telephone technology, and his XYZ force, but he also sketched out ideas for a miscellany of other inventions.

The couple returned north in April and settled into their new home in Llewellyn Park in West Orange, New Jersey. Soon after, Edison moved his laboratory to the Harrison lamp factory and undertook an extensive series of experiments to improve the lamp and its manufacture. In the fall, recognizing the competitive threat from George Westinghouse's new high-voltage alternating current (AC) electrical system, he resumed experiments on improved electrical distribution. At the same time, the development of the graphophone, a wax-recording phonograph devised at Alexander Graham Bell's Volta Laboratory, pushed Edison to resume work on the phonograph. By the spring of 1887, Edison decided to build a much larger laboratory near his home in West Orange where he could experiment simultaneously on these technologies and begin exploring other ideas from his Fort Myers notebooks.

THE WEST ORANGE LABORATORY

Edison's decision to build his new facility may also have been spurred by accounts of rival electrical inventor Edward Weston's new laboratory in Newark, which one journalist called "the most complete private lab in the country" (Israel 1998, 260). Edison aimed to build "the best equipped & largest Laboratory extant," with facilities "incomparably superior to any other for rapid & cheap development of an invention." It cost $180,000 (nearly $6 million today), but Edison could justifiably claim that "there is no similar institution in existence" (*TAEB* 8:784). There he could apply the processes of research and development he had used at Menlo Park on a much larger scale to develop a wide variety of technologies— especially electric light and power, sound recording, motion pictures, iron-ore milling, primary and storage batteries, and Portland cement and chemical manufacture. Factories to manufacture his inventions would spring up around the laboratory. The West Orange complex helped lay the groundwork for modern industrial research, while Edison's success at moving inventions from laboratory to market provided a model for the modern process of innovation.

Edison's high ambitions for the new facility are evident in a four-page project list and in the size of his staff, which numbered between eighty and 100 men. To keep track of them all, each employee had a weekly time sheet on which was recorded his job title, hours worked, pay rate, wages earned, and often the project accounts to which his wages were charged. Edison hoped to build an adjacent "great Industrial works" but could not find a partner to share the laboratory's ongoing expenses and prospective profits (*TAEB* 8:785). Unable to bear the costs alone, he drastically cut the staff in April 1889 and focused his research on a few major projects.

THE CURRENT WAR

In its early years, the West Orange laboratory served as the research and development center for Edison electric lighting companies. His electrical laboratory, supervised by Arthur Kennelly, undertook much of this work. But even as Edison was opening his new laboratory, his preeminent position in the American electric power industry was threatened by AC systems that, light for light and mile for mile, were cheaper to build and operate than his direct current (DC) system. Slow to recognize the threat, Edison initially responded with a burst of innovation, including improvements to his lamps, to make DC more competitive. Unable to find a technological advantage, he focused instead on a vitriolic campaign against George Westinghouse, the most visible AC promoter, that entangled him in debates over the role of electricity in euthanasia and capital punishment. The multiyear contest for market dominance between DC and AC systems, represented by Edison and Westinghouse, has become known as the War of the Currents.

As part of the effort to compete against Westinghouse and other rival companies, Edison and Henry Villard, an Edison Electric investor who funded Edison's electric railway experiments, developed a plan to recapitalize the Edison lighting interests. Its centerpiece was the consolidation of the independent manufacturing plants with Edison Electric into a new Edison General Electric Company (EGE). EGE also took over the Sprague Electric Railway & Motor Company, but Edison consistently opposed efforts to merge EGE with Thomson-Houston Electric, the third major electric manufacturer and (with Sprague) a leader in railways and motors. Electric rail lines, which were starting to reshape cities large and small, and industrial motors beckoned as lucrative new markets for EGE and its rivals. Nonetheless, Edison's resistance to AC put EGE at a disadvantage. In 1892, investors led by J. P. Morgan forced EGE into a merger with Thomson-Houston to form a new entity, the General Electric Company, which had neither Edison's name on the outside nor his key allies inside. DC did not disappear, but Edison lost the war and with it the services of Samuel Insull, his longtime associate and business manager, who left New York to start a new career in Chicago.

THE "PERFECTED" PHONOGRAPH

Although Edison had been working on a new wax-cylinder phonograph for several months before opening his new laboratory, he now had the resources to attack the problem more systematically. He put one team of researchers to work developing materials for the records, another on duplicating recordings, and other teams on the phonograph's mechanical construction, its motor and battery, and recording and playback devices. In addition, Charles Batchelor headed work on a talking doll with a miniature phonograph inside. Edison erected a factory next to the laboratory to manufacture the phonograph and the doll. He also built a plant in nearby Silver Lake to make his new primary battery and wax for phonograph records.

When Edison brought out his so-called perfected phonograph in mid-1888, he sent one to his longtime London agent, George Gouraud, to whom he also assigned phonograph marketing rights in Europe and most of the rest of the world. He then entertained a proposal brought to him by personal attorney John Tomlinson and close friend Ezra Gilliland (whom he had made general sales agent for the United States and Canada). Industrialist Jesse Lippincott offered half a million dollars for Edison's domestic phonograph rights, which he intended to combine with the rival graphophone in a single entity, the North American Phonograph Company. Needing money to run the laboratory, Edison accepted Lippincott's terms (reserving to himself the rights to manufacture the phonograph and make duplicate recordings). Unbeknownst to him, however, Gilliland and Tomlinson had made private arrangements with Lippincott for their own benefit. Edison became incensed when he discovered these deals and permanently severed relations with both men.

North American leased phonographs and graphophones to regional subcompanies that in turn rented the machines to local businessmen for dictation. Although dictation was seen as the primary market for sound recording, Edison also expected his phonograph to be used for recorded music. Indeed, the laboratory recorded nearly 900 musical cylinders for playback at the Paris Universal Exposition of 1889, where the phonograph was the sensation of Edison's large exhibit. Edison and Mina traveled to Paris for the exposition, where he was feted everywhere he went. By the end of 1890, Edison phonographs were being used with a nickel-in-slot device that enabled users to listen to a recording for five cents. Over the next few years, this commercial use surpassed dictation as the primary market for sound recording.

MOTION PICTURES

As the phonograph became an entertainment device, it was joined by another Edison invention, motion pictures, on which Edison had begun experimenting in October 1888. He assigned the project to experimenter William K. L. Dickson, who was also a photographer. Because Dickson doubled as the primary experimenter on Edison's ore milling project,

motion pictures often took a back seat. Thus, it was not until 1891 that Edison applied for his basic patents on a camera (the "kinetograph") and a peephole viewing device (the "kinetoscope"). And it took until the middle of 1894 for the first kinetoscope parlors to open to the public. Although Edison, with Dickson's help, was the first to commercialize motion pictures, the peephole kinetoscope proved to be little more than a novelty. Other inventors soon developed projectors and more portable cameras that helped the new industry take off. After Dickson left to form another company, Edison placed his business under the direction of the Edison Manufacturing Company, which manufactured his primary battery and phonograph wax. Although Edison's name appeared on every film produced by the company, he had little to do with the motion picture business.

MATERIALS PROCESSING (IRON AND CEMENT)

Although today Dickson is often credited as the coinventor of motion pictures, his only joint patent with Edison was for an improved electromagnetic separator for refining low-grade iron ores. That patent, filed in January 1890, marks the start of what became Edison's decade-long effort to revive eastern iron mines and furnaces. The device used powerful electromagnets to pull magnetic material from a mass of crushed rock or sand dropping into a stream or moving on a belt past the magnet. Edison had developed his first magnetic ore separator in 1880 to extract iron particles from black sand, but by 1887, he focused on separating and purifying the low-grade ores found in played-out mines in and around New Jersey. After forming the New Jersey and Pennsylvania Concentrating Works in 1890, he acquired mining properties around Ogdensburg, New Jersey, where he began building an integrated processing plant. There he worked for most of the next decade, coming home only on weekends. He designed, tested, and redesigned semiautomated systems for mining, crushing, separating, and concentrating the ore and molding it into briquettes. Edison poured more than $2 million of his own money into this venture but was forced to abandon it after the expansion of mining in the iron-ore–rich Mesabi Range of the Great Lakes region made his concentrated low-grade ore too expensive, especially as the steel industry shifted to the Midwest.

Edison had some success selling waste sand from his ore milling project to cement manufacturers. In 1899, he decided to investigate transferring this technology (especially his system for crushing rock) to the production of Portland cement. During the next few years, Edison made other improvements in cement manufacture, the most important of which was a long rotary kiln that he used at an automated plant he built near Stewartsville (in western New Jersey) and licensed to other manufacturers. He also designed a system for building inexpensive cement houses that he freely licensed to builders; several such houses were constructed in the 1910s.

SOUND RECORDING INDUSTRY

As Edison's ore milling venture petered out in the late 1890s, the phonograph business finally took off. Edison had taken control of North American Phonograph and reorganized his phonograph business as National Phonograph, one of the few companies that did not bear his name. By this time, the market was driven by the development of cheap machines run by spring motors. Such instruments were better suited to a mass market for home entertainment than his expensive electric model, which required consumers to replenish the chemicals and electrodes in the primary battery. After adopting and improving spring motors for his phonographs, Edison and his staff finally perfected a method of molding duplicate records at the turn of the century. Business boomed, and Edison's company became the clear leader in the field.

STORAGE BATTERIES

Edison devoted little attention to his phonograph business during the first years of the twentieth century. Instead, he invested the sizable income it created into the cement works and his effort to invent a better battery for electric automobiles. Aware of the weight problems with lead-acid batteries, Edison decided to try alkaline electrolytes, such as those used in his primary battery, to develop a lightweight and long-lasting cell. However, it took him a decade to develop a commercially viable iron-nickel battery, by which time internal combustion engines prevailed. Edison did find an extensive market for his battery in a variety of industrial uses, and it became the most successful product of his later life.

THOMAS A. EDISON INCORPORATED

Edison cylinder phonographs and records dominated the industry until the first decade of the new century, when they began to face strong competition from the Victor Talking Machine Company and its disc gramophone. While the Panic of 1907 hurt both Victor and Edison's National Phonograph Company, Victor emerged from the crisis as the industry leader. Its record catalog offered a greater variety of artists; the Red Label catalog, in particular, featured prominent opera stars like Enrico Caruso. These records appealed to affluent urban consumers whom the economic downturn affected less severely than National Phonograph's small-town and rural customer base. Victor's instruments also appealed more strongly to middle-class consumers. In 1906, Victor began offering the Victrola, which made the record player an elegant piece of furniture by hiding the machine and its horn in an attractive cabinet.

Edison's efforts to regain industry leadership began with the development of his own internal-horn phonograph, the Amberola. It entered the market in 1909 and could play both the standard two-minute cylinders and the new four-minute versions. In 1911, at the urging of corporate attorney Frank Dyer, he organized Thomas A. Edison Incorporated as a multidivisional corporation that included separate units for consumer and business phonographs, recordings, and record production. The same year, he finally began offering a disc phonograph designed to surpass the Victrola in sound quality. Although many consumers agreed that the Edison disc sounded better, they complained about the company's choices of songs and artists, which Edison—despite his increasing deafness and age—had assumed responsibility for selecting. The phonograph business declined steadily until Edison abandoned it in 1929.

WORLD WAR I

With the outbreak of war in Europe in August 1914, there was growing concern that the United States might be drawn into the conflict. In mid-1915, following the U-boat attack on the steamship *Lusitania*, Edison proposed a plan of national preparedness that led Secretary of the Navy Josephus Daniels to establish the Naval Consulting Board (NCB). Daniels asked Edison to head the NCB and help choose its members from among the nation's leading inventors and engineers. Several NCB members (including Edison) conducted their own research in addition to reviewing inventions for the navy. Edison presented more than fifty original ideas, none of which was adopted. The navy also rejected his proposal for the Naval Research Laboratory it established after the war. Edison was more successful in his campaign to overcome shortages of chemicals he had previously obtained from Germany and England. After studying the problem, he quickly erected several manufacturing plants and became a major chemical supplier to American industries, European allies, and Japan.

EDISON'S LAST CAMPAIGN

The return of peace did not alleviate concerns about the British monopoly on rubber production and the possibility that another war could disrupt supplies of this vital natural commodity. The United States was particularly vulnerable because it consumed more than 70 percent of the world's rubber, much of it in the rapidly expanding automobile industry. Edison, who needed rubber for insulation in his storage batteries, joined with automobile manufacturer Henry Ford and tire magnate Harvey Firestone to develop an alternative supply in

case of national emergency. In 1927, they established the Edison Botanic Research Corporation, whose purpose was to fund Edison's search for plants that could be cultivated on American soil that would yield sufficient latex at a cost competitive with imported rubber. Edison identified a strain of goldenrod as particularly promising, but he died before completing the project.

EDISON'S LEGACY

From the moment he became the Wizard of Menlo Park in 1878, following his invention of the phonograph, Edison lived in the glaring light of modern celebrity, a spotlight he welcomed and sometimes directed. Reporters followed his inventive activities and sought his opinion on subjects ranging from the technologies of the future to questions of diet and the existence of God. He became a near-universal symbol of American inventiveness and industrial progress. It was for this reason that he received a Congressional Gold Medal in 1928 and the next year—the fiftieth anniversary of the electric lamp—was designated as Light's Golden Jubilee. His friend Henry Ford staged a ceremony attended by President and Mrs. Hoover that was broadcast across the country. The acclaim was not universal. Scientific research had advanced well beyond Edison's appreciation or comprehension, and its practitioners—among whom he had once moved easily—for a time blocked his election to the National Academy. Indeed, "research" had largely replaced "invention" as a desired process. But Edison remained a popular icon. Newspapers followed his declining health, and his death on 18 October 1931 was the lead headline in newspapers around the world. Heads of state, government bodies, business and civic organizations, citizens, and schoolchildren sent condolences. When Edison was buried three days later, President Hoover asked the nation to dim its lights in his honor, and radio broadcasts observed a moment of silence to pay tribute.

Edison's death seemed to mark the passing of an era. Although his own laboratories helped lay the groundwork for the modern industrial research and development lab, Edison had remained, as one of his Menlo Park experimenters described him, the "sole directing mind" (Israel 1998, 195). That individualistic model was increasingly outdated, overtaken by some of the very changes Edison had helped to set in motion. By the 1920s, as he retired from inventive work, he failed to replace himself with a director of corporate research who could help innovate his company's future. Instead, his laboratory became little more than a product improvement lab for existing products. As a consequence, Edison was recalled at his death as the last of the lone inventors rather than celebrated as the founder of American industrial research.

The Dictionary

ACHESON, EDWARD GOODRICH (1856–1931).

Pursuing an interest in electricity, Acheson started as a draftsman in Edison's laboratory in **Menlo Park, New Jersey**, in September 1880 and soon became an experimenter. Edison contracted with him in early 1881 to create and manufacture lamp filaments from pressed plumbago, a form of graphite. Acheson did so but renounced the lucrative arrangement when the filaments proved short lived. He subsequently assisted **Charles Batchelor** at the **Paris International Electrical Exposition (1881)**, established a lamp factory on that city's outskirts, and installed several isolated lighting plants in Europe. Acheson left the Edison ventures in 1882 but returned in early 1884 to work in Edison's **New York City** laboratory. Leaving again after just a few months, he built an outstanding career as an independent inventor and engineer, securing his legacy in 1890–1891 with the invention of carborundum, an industrial abrasive.

ACOUSTIC TELEGRAPH.

Systems of acoustic telegraphy employed a series of tuning forks or reeds to transmit multiple signals at different frequencies at the same time over a single wire. The development of acoustic telegraphs began in earnest in 1875 as alternatives to the **duplex** and **quadruplex** systems of **Western Union**. Inventors working on acoustic telegraphy included Elisha Gray, **Alexander Graham Bell**, and Edison, with Gray giving the first public demonstration. Edison likely saw it in **New York City** in August 1875 and mentioned it in his first acoustic telegraphy patent caveat.

Edison's acoustic telegraph experiments led to the discovery of what he called etheric force, one of several supposed **unknown natural forces**. He also experimented with an alternative system he called acoustic transfer telegraphy. It used tuning forks to switch the circuit rapidly among several sets of standard Morse telegraph instruments as a method for time-sharing of the transmission line. Edison's work on acoustic and acoustic transfer telegraphy concluded by the end of 1876 as he focused on improvements to the telephone, which had grown out of the acoustic telegraph research of Gray and Bell.

ADAMS, JAMES (ca. 1845–1879).

Little is known of Adams, who was one of Edison's principal assistants in experiments on **telegraphy**, the telephone, and the **electric pen**. Born in Scotland, Adams apprenticed as a cabin boy and became a sailor. Edison first employed Adams in 1874 as a Brooklyn agent for Edison's inductorium (a medical shock device) and other electrical equipment. Adams witnessed a notebook entry for Edison in August 1874 but was not a regular member of the experimental staff in **Newark, New Jersey**, until May 1875. He remained on the staff when Edison moved into the new **laboratory** in **Menlo Park, New Jersey**, early the next year. His contributions to the development of Edison's electric pen and telephone earned Adams a percentage of patent royalties from those inventions. In March 1878, Edison sent him to London as his technical representative to promote the telephone in Britain and

France. Adams fell seriously ill in February 1879 and died in London three months later. Edison continued to pay royalties to Adams's widow Ellen until the patents expired in 1888.

ALTERNATING CURRENT (AC). In an AC circuit, the flow of electrical energy reverses direction at regular intervals, typically many times each second. AC is the natural output of rotating generators without commutators or other devices to "straighten" the flow into **direct current (DC)**. There was little commercial use for AC until the mid-1880s, when the development of AC meters, motors, and transformers made it a practical alternative to DC and led to the **War of the Currents**. The lack of commutating devices made AC motors and generators simpler than their DC counterparts. A bigger advantage for AC was the ability to easily raise or lower voltage (electrical "pressure") by transformers. Sending power at high voltage over long distances with little loss and then distributing the power at lower voltage for lamps and motors greatly extended the area that a generating station could serve economically. The high voltages at which AC was often distributed did pose safety risks, however. The fluctuating magnetic fields of AC seemed to Edison a waste of energy and made it harder to model the circuits mathematically.

AMERICAN ASSOCIATION FOR THE ADVANCEMENT OF SCIENCE (AAAS). Founded in 1848 at the Academy of Natural Sciences in Philadelphia, the AAAS was the first permanent organization to promote the development of science and engineering at the national level. **George Barker** formally presented Edison as a new member at the 1878 annual meeting in St. Louis, where Edison presented a paper on the **tasimeter**. That paper was to be the only one that Edison—a reluctant public speaker—personally read to an AAAS gathering. He demonstrated his telephone at the next year's annual meeting in Saratoga, New York, but Barker and **Francis Upton** made presentations on his behalf about resonant tuning forks and electric light experiments, and Barker read Edison's paper on the **pyromagnetic generator and motor** at the

1887 gathering. **Henry Rowland** used the 1883 annual meeting to attack applied science and technology. In a thinly veiled reference to Edison, he objected to calling "telegraphs, electric lights, and such conveniences, by the name of science" (Israel 1998, 465). As Rowland intended, his "Plea for Pure Science" came to symbolize a divergence between the practical work of Edison, however revered by the public, and that of academic scientists. *See also* NATIONAL ACADEMY OF SCIENCES (NAS); SCIENCE (EDISON AND).

AMERICAN BELL TELEPHONE COMPANY. This **Boston**-based company owned the principal telephone patents in the United States and created an effective monopoly on telephone service by the mid-1880s. Formed in 1880 by reorganizing the National Bell Telephone Company (which existed to commercialize **Alexander Graham Bell**'s invention), American Bell also acquired the telephone patents of Edison and Elisha Gray from **Western Union Telegraph Company**. Edison's friend **Ezra Gilliland** ran American Bell's experimental shop, and in 1884, the company agreed to contract with Edison for research aimed at improving its new long-distance telephone service. Although that arrangement was never finalized, Edison did some work on American Bell's behalf. The company also assumed his earlier telephone contract with Western Union and paid him $6,000 a year through 1907.

AMERICAN DISTRICT TELEGRAPH COMPANY (ADT). In 1871, **Edward Calahan** invented a "district and fire-alarm telegraph" to provide "warning in cases of fires, burglars, or accidents" (*TAEB* 1:411 n. 2). Together with associates from the **Gold and Stock Telegraph Company**, Calahan incorporated the American District Telegraph Company in October 1871. The company established offices in several Brooklyn and **New York City** neighborhoods and placed small transmitting devices in private homes. Subscribers could signal for police, firemen, or physicians through the call boxes, and the company's distinctively clad messenger boys would respond from its offices. In January 1872, Edison made extensive notes regarding improvements to the ADT system, filing a patent caveat soon thereafter that he assigned to ADT. (His patent application based on that caveat was rejected.) Sensing the possibilities in the business, Edison cofounded the Domestic Telegraph Company in March 1874 as a direct competitor to ADT. By the end of the year, Domestic had lines in New York; **Newark, New Jersey**; and Canada using instruments made by Edison and **Joseph Murray**. Domestic Telegraph was acquired in 1876 by **Jay Gould**'s **Atlantic and Pacific Telegraph Company**, and its system fell into disuse. However, ADT continued to thrive with more than 2,000 subscribers. Edison himself attested to the ubiquity and reliability of its services when, in the fall of 1883, he asked to have an ADT call box put in his residence at the Clarendon Hotel in New York City. **Western Union** aggregated the local ADT affiliates into a single national company in 1901. Today, ADT is a multi-billion-dollar security company based in Boca Raton, Florida.

AMERICAN INSTITUTE OF ELECTRICAL ENGINEERS (AIEE). The AIEE was founded in 1884 by prominent figures in the electrical industry, including Edison, who was elected one its vice presidents. Edison's role in the AIEE was largely honorary, and he generally did not participate in its meetings or activities. Other individuals from the Edison electrical companies or his **laboratory** did participate more actively, including giving papers. Two notable papers were closely associated with Edison's work. The first, "Some Notes on Incandescent Lamps" by Edwin Houston, dealt with the **Edison Effect** and was delivered at the inaugural meeting in 1884. The second was a technical paper for the May 1901 meeting about the new alkaline **storage battery** Edison was developing for electric automobiles. Edison drafted it, but because of his aversion to public speaking, he had his former assistant **Arthur Kennelly** present it. The AIEE recognized Edison's contributions to electrical engineering by establishing the Edison Medal as its highest award in 1908. (It was first awarded the following year.) Following the 1963 merger of the AIEE with the Institute of Radio Engineers

to form the Institute of Electrical and Electronic Engineers (IEEE), the IEEE adopted the Edison Medal as its principal honor for "a career of meritorious achievement in electrical science, electrical engineering, or the electrical arts."

AMERICAN MUTOSCOPE AND BIOGRAPH COMPANY. In 1895, Bernard Koopman, Henry Marvin, Herman Casler, and former Edison employee and film pioneer **William K. L. Dickson** founded the K.M.C.D. Company in **New York City** to produce flip-card **motion pictures** for the Mutoscope, a peephole viewing machine. However, the company soon began to produce films for projection in competition with the **Edison Manufacturing Company** and changed its name to the American Mutoscope and Biograph Company. Edison confronted the new concern with a patent lawsuit to prevent it from issuing films for projection, but in 1902, American Mutoscope and Biograph prevailed in court. Initially, its documentary and entertainment films were shorts of no more than a minute or two, but with the release of *The Battle of the Yalu* in 1904, the company began to compete with the Edison Manufacturing Company in the production and distribution of longer features. In 1908, American Mutoscope and Biograph joined with Edison Manufacturing to form the **Motion Picture Patents Company**, a patent-pooling and licensing firm. American Mutoscope and Biograph shortened its name to the Biograph Company in 1909 to signal its commitment to making films for projection. After ceasing operations in 1928, the company's assets were sold to Consolidated Film Industries. Directors D. W. Griffith and Mack Sennett and actors Lillian Gish and Blanche Sweet began their film careers with American Mutoscope and Biograph.

AMERICAN TELEGRAPH WORKS (ATW). Edison and **George Harrington** founded ATW, a partnership to manufacture telegraph instruments, in October 1870. Harrington supplied $6,000 of the shop's initial capital, with Edison putting in $3,000 and machinery and tools from the Newark Telegraph Works (his partnership with **William Unger**). ATW was immediately one of the larger electrical manufacturing firms in the country, prompting Edison to tell his parents that "I am now-what 'you' Democrats call a 'Bloated Eastern Manufacturer'" (*TAEB* 1:212). Over the next year, ATW would make quotation instruments for the cotton trade, perforators, and similar items. Edison chafed at his bill-paying and payroll responsibilities, complaining to Harrington that "if I keep on in this way [much] longer I shall be completely broken down in health & mind" (*TAEB* 1:308). Edison may also have resented Harrington's replacement of a pliant supervisor with Spencer Clark, who did not indulge Edison's penchant for refining designs during the course of manufacture (and who would later accuse Edison of stealing credit for **automatic telegraphy**). Edison left ATW in October 1871, taking workmen (including **Charles Batchelor**) and machinery to Edison & Unger, successor to the Newark Telegraph Works. The firm of Edison & Murray subsequently acquired the business and remaining equipment of ATW, which ceased operations in early 1873. *See also* MURRAY, JOSEPH THOMAS.

ARMINGTON & SIMS ENGINE COMPANY. This firm was a major supplier of high-speed stationary steam engines for Edison and his electric lighting companies. Initially a partnership of Pardon Armington and Gardiner Sims in Lawrence, Massachusetts, it incorporated in Rhode Island in 1882 as the Armington & Sims Company and moved to Providence. It expanded with Edison's encouragement and, in 1883, incorporated again under its present name. In early 1881, Edison was looking for engines to drive 125-horsepower dynamos directly (without belts or gears) in urban central stations. Armington & Sims had an economical engine whose sensitive governor and rigid frame partially filled Edison's needs, and they adapted it to the higher speed (350 revolutions per minute) required for the direct connection. Unable to acquire the engines quickly enough, Edison negotiated the right to build them himself (though he never did so). Armington & Sims remained the engine of choice for large Edison central stations in the United States and abroad as well as for smaller plants

built by the **Edison Construction Department** and isolated lighting plants.

ARTIFICIAL MATERIALS. Edison first experimented with artificial materials in 1875 to develop an ivory substitute less expensive than celluloid, the first commercial plastic material (made in **Newark, New Jersey**, by the Celluloid Manufacturing Company). In 1879, he tried unsuccessfully to develop a process for making a compound that looked like gold. During the 1880s, he sought to create artificial alternatives to ivory, silk, mother of pearl, and hard rubber. The experiments were unsuccessful, but Edison included the idea of such replacements for expensive materials in the notes he gave to **George Parsons Lathrop** for a proposed science fiction novel.

Edison made his most sustained research into artificial materials while attempting to develop better **electric lamp** filaments and phonograph records in the 1880s. His filament work failed to produce a material better than natural bamboo, though inventors **Joseph Swan** and **Edward Weston** independently created squirted filaments made from artificial materials—filaments that became standard in the lighting industry. Edison's experiments with phonograph record materials, led by chemist **Jonas Aylsworth**, were more successful. Their synthesis of waxlike compounds for cylinders helped Edison's company dominate the early sound-recording industry. When longer-playing disc records threatened the company's position in the twentieth century, Aylsworth developed Condensite, a phenol plastic, for Edison's own disc recordings.

ASTRONOMY. Edison maintained a lifelong interest in astronomy and sometimes engaged in speculation about it. However, he did not conduct serious astronomical research except in the late 1870s while developing and testing the **tasimeter**. In that period, he corresponded with astronomer Henry Draper about the discovery of solar oxygen and with astronomer and physicist Samuel Langley on ways to measure the heat of stellar spectra. In 1878, Edison borrowed a telescope to observe the transit of Mercury across the sun. The same year, he famously joined Draper's solar eclipse expedition to Rawlins, Wyoming, where he tried out the tasimeter—with mixed results. Edison made a series of speculative notebook entries in 1886, some of which touched on celestial mechanics. He posited gravitation as an electromagnetic phenomenon, suggesting that Earth and the sun act on one another as powerful magnets. Alternatively, he conjectured that since each atom must have some polarity, gravitation might be explained as a "mutual attraction of all the atoms" (*TAEB* 8:483). A brief but intense period of reading works on astronomy that same year led him to muse about the electromagnetic origins of sunspots and comets. During his 1889 trip to France, he visited astronomer Jules Janssen at the observatory at Meudon, where he inspected Janssen's gyroscope. He may also have seen the photographic revolver that Janssen used to record the motion of heavenly bodies by automatically taking forty-eight photographs in succession. Edison periodically returned to speculations on these subjects later in life. *See also* ELECTROMAGNETISM; MATTER (EDISON'S CONCEPTIONS OF); UNKNOWN NATURAL FORCES.

ATLANTIC AND PACIFIC TELEGRAPH COMPANY. Organized in 1867, the relatively small Atlantic and Pacific struggled financially. In 1874, financier **Jay Gould** acquired it as part of his effort to challenge the dominance of **Western Union**. Atlantic and Pacific consequently took over the **Automatic Telegraph Company**. Edison became Atlantic and Pacific's electrician, a position he held for about six months as he worked to equip its lines with his **automatic telegraph** system. Gould's initial challenge to the industry-leading Western Union failed, but the price war launched by Atlantic and Pacific weakened the larger company and later enabled Gould to gain control of it.

AUTOMATIC PHONOGRAPH EXHIBITION COMPANY (APEC). Seeking to capitalize on popular interest in coin-operated equipment of all kinds and make the most of disappointing **phonograph** revenues, Charles Cheever and Felix Gottschalk, entrepreneurs with the New York City–based Metropolitan Phonograph Company, organized APEC in New York in February 1890. It was intended to make, lease, and sell coin-operated phonographs and related equipment, specifically Albert Keller's nickel-in-slot device. Keller had worked on such an attachment in 1887 with the encouragement of **Ezra Gilliland** while both men were employed by Edison. Keller completed the project in late 1889, by which time he was working for the family firm of Gilliland Electric Company and Gilliland himself had been cast out of the Edison orbit. APEC ordered 500 Keller instruments from Gilliland Electric, each to be enclosed in a wooden cabinet.

Edison had heretofore taken only a cursory interest in such machinery, but he became alarmed that the now-despised Gilliland might wring some personal profit from the phonograph. APEC was not tied exclusively to Keller's invention, and Edison exerted what leverage he had to freeze Gilliland out of the business. Thus, in April 1890, he approved an arrangement giving APEC control of the national market for coin-operated phonograph devices but reserving their manufacture to the **Edison Phonograph Works**, which also had effective veto power over the choice of cabinetmaker. Yet APEC kept doing business with Gilliland under a provision that allowed it to accept devices from Gilliland Electric until the Edison Phonograph Works could make them at a steady pace. **John Ott** worked sporadically on a new nickel-in-slot device, but Edison did not at first give the project much urgency, and preparations at the Phonograph Works languished.

APEC eventually controlled the patents of Edison, nickel-in-slot pioneer Louis Glass, and others. However, it struggled due to the ambivalence of phonograph companies, competition from a proliferation of such devices, and the sales practices of the **North American Phonograph Company**. (APEC had an exclusive arrangement with North American but was hurt when the latter in 1890 allowed phonographs to be sold rather than leased, affording buyers their choice of coin attachments.) APEC went into receivership in 1894 and was dissolved before it could benefit from an upsurge in the popularity of nickel-in-slot phonographs.

AUTOMATIC TELEGRAPH. This high-speed telegraph system used a perforating machine to punch holes representing Morse code into a strip of paper. The operator then fed the punched strip into a transmitter, where a metal stylus or roller would make electrical contact with a revolving drum when the perforations passed between them, thus closing the circuit. At the other end of the line, the intermittent electrical signals charged a metal stylus touching chemically treated paper, causing the chemicals to decompose and leave images of the Morse dots and dashes. Automatic chemical-recording telegraphs were based on the original 1846 design of Alexander Bain. Alternative automatic systems, notably one devised by Englishman Charles Wheatstone and used extensively in Great Britain, employed ink recorders that were much slower than chemical recorders. The **Automatic Telegraph Company** employed Edison in 1870 to make improvements in its automatic telegraph system (patented by British inventor George Little). Edison so substantially modified the company's system over the next three years that it became identified with his name.

AUTOMATIC TELEGRAPH COMPANY (ATC). Former assistant secretary of the treasury **George Harrington** and several associates formed ATC in November 1870, with Harrington as president. It subsequently acquired rights to George Little's **automatic telegraph** from journalist **Daniel Craig** and the National Telegraph Company, which Craig had engaged to develop Little's system for transmitting press copy. Craig also contracted with Edison to improve Little's system. Impressed by Edison's work, Harrington gave him the necessary resources to design his own system. Harrington also became a partner in the **American Telegraph Works**, which developed and manufactured Edison's automatic telegraph equipment. ATC secretary **Josiah Reiff** provided Edison with a $2,000 annual salary and funded much of his work for ATC. By the time ATC began providing commercial service in 1872, it was using Edison's automatic telegraph improvements. However, the company struggled financially, and in December 1874, its shareholders agreed to a takeover by the **Atlantic and Pacific Telegraph Company** of **Jay Gould.** Still unpaid a year later, they established a new company, American Automatic Telegraph Company, to which Edison assigned his patent rights in automatic telegraphy. This effort to force a settlement with Gould and A&P did not succeed.

AYLSWORTH, JONAS WALTER (1868–1916). A longtime and valued collaborator with Edison in chemical matters, Aylsworth studied a year at Purdue University (near his hometown) before joining the **laboratory** staff in **West Orange, New Jersey,** in late 1887 or early 1888. He quickly took a lead role in developing new artificial wax compounds for phonograph cylinders and later organized wax production for the **Edison Phonograph Works.** From 1891 to 1894, he oversaw production of cellulose lamp filaments for the **Edison Lamp Company,** first at a secret facility in Jersey City and then at the main factory in East Newark (Harrison), New Jersey. In 1894, Aylsworth and F. E. Jackson formed a partnership to manufacture lamp filaments (and later Edison's fluoroscope), even as Aylsworth continued working five days a week at Edison's laboratory.

After dissolving the partnership with Jackson in 1898, Aylsworth worked on a contract basis for Edison, chiefly in a laboratory he built for himself at the back of his house in East Orange. In addition to his intensive work on phonograph waxes and lamp filaments, Aylsworth assisted Edison in ore milling, storage batteries, and rubber research. While attempting to produce a hard compound for disc records, he improved the phenol resins developed by Leo Baekeland. He called his new purer resin Condensite, and in 1910, he founded the Bloomfield, New Jersey–based Condensite Company of America, which licensed the polymer for Edison Diamond Disc records. Aylsworth received more than 100 patents (most related to Condensite) in his lifetime, and Edison called him "one of the best empirical experimenters I ever have known" (Jeffrey 2008, 40).

B

BANKER, JAMES HOPSON (1827–1885). An important supporter of Edison's lighting enterprises, Banker made his fortune in his family's ship-chandlery business in New York City. He retired in 1869 to focus on his work as a banker and broker, mainly as vice president of the Bank of New York. He became a good friend of Cornelius Vanderbilt and was closely associated with the Commodore's New York railroad ventures. Banker joined the **Western Union Telegraph Company** board of directors in 1870 after Vanderbilt acquired a controlling interest in the company; he also served on the executive committee of **Gold and Stock Telegraph Company.** In 1878, Banker joined other Western Union investors to incorporate the **Edison Electric Light Company** and became a member of its executive committee. He had a key role in organizing the Edison Electric Light Company of Europe in 1880. He also helped create the Edison Telephone Company of Europe.

BARKER, GEORGE FREDERICK (1835–1910). Barker was a prominent chemist and physicist with whom Edison had a long, useful, and (for the most part) friendly relationship. Barker was professor of physics at the University of Pennsylvania from 1873 to 1900, a member of the **National Academy of Sciences** (NAS), and president of the **American Association for the Advancement of Science** in 1879. His association with Edison began in 1874, when Barker saw the **electromotograph** at the Franklin Institute and arranged for Edison to discuss it at the NAS meeting in Phil-adelphia. Four years later, Barker spoke on Edison's behalf when the inventor exhibited his **phonograph** and **carbon-button telephone transmitter** to the NAS at the Smithsonian Institution. Over the next few years, he served as the inventor's liaison to the scientific community, helping him draft and deliver papers. Barker invited Edison on a solar eclipse expedition to Wyoming in 1878, for which Edison hoped to use his **tasimeter**. While traveling, the two men mused about the possibilities of electric light and power, and soon after their return, Barker arranged for Edison to see the lights and generator being made by electrical manufacturer William Wallace at Ansonia, Connecticut. The Ansonia visit filled Edison with premature confidence and marked the start of his full-time effort to create an electric alternative to indoor **gas lighting**. In 1880, Barker and fellow scientist **Henry Rowland** rigorously tested the efficiency of Edison's carbon-filament **electric lamp** and reported favorably. However, later that year, the well-respected Barker was quoted in print praising the rival lamp of **Hiram Maxim**. Pressed by Edison, Barker admitted being impressed by the Maxim lamps but denied the quotes attributed to him. Barker subsequently served the **Edison Electric Light Company** on retainer and accepted Edison's personal gift of stock in the company, but the relationship between the two men, while cordial, was never the same.

BATCHELOR, CHARLES (1845–1910). A proverbial and literal right-hand man, Batchelor worked side by side with Edison through many of the inventor's most creative years. Sometimes described as a skilled machinist, Batchelor possessed rare technical skills, especially a facility for delicate work. He was also a creative problem-solver, and his ability to systematize anything at hand, from intricate experiments to production processes, proved invaluable. He and Edison regarded each other as friends.

Batchelor was born in London and raised near Manchester, England. He did well in school and entered Manchester's textile mills, probably as an apprentice. His talents were evident, and in 1870, the J. P. Coates Company (famous thread makers) deputized him to help set up a factory in **Newark, New Jersey**. By October of that year, Batchelor was working at the **American Telegraph Works**. That marked the start of his long association with Edison, whom he followed to a series of Newark shops and laboratories and then (in 1876) to **Menlo Park, New Jersey**. Batchelor was essential in systematically trying out Edison's ideas—and adding his own—while methodically recording the results. He collaborated with Edison on every major project at his laboratories in Newark and Menlo Park. (An incomplete list would include the **electric pen, electromotograph telephone receiver, carbon-button telephone transmitter, phonograph**, incandescent **electric lamp, and electric generator**.) After setting up Edison's first lamp factory at Menlo Park in 1880, Batchelor went to France to begin making lamps and other electric light components for European markets (and to satisfy requirements of French patent law). Edison recalled him in 1883 to oversee expansion of the **Edison Machine Works**, and Batchelor followed the Works to Schenectady, New York. He helped manage the construction and start-up of the **laboratory** in **West Orange, New Jersey**. Around 1890, he joined Edison's iron-**ore milling** quest near Ogdensburg, New Jersey. He assisted at the mine until Edison closed it at the turn of the century. Batchelor had about ten years of retirement, during which he stayed in contact with Edison. Enabled by Edison-related royalties and stock dividends, Batchelor traveled extensively with his wife Emma and their two daughters until his death.

BATTERIES (PRIMARY). Primary batteries use a chemical reaction to create **direct current (DC)** electricity. In Edison's lifetime, the reaction was typically made by electrodes of copper, zinc, or similar metals in a liquid acid. Unlike **storage batteries**, primary batteries cannot be recharged; when depleted, their reactive materials must be renewed or replaced. Primary batteries were indispensable for powering telegraph and telephone systems, and Edison gained an early familiarity with them as a telegrapher. Later, he and his **laboratory** assistants sometimes used analogies with primary batteries to understand electrical behavior in novel devices like **electric generators**, lighting circuits, the **railway** ("grasshopper") **telegraph**, and converters of **alternating current** to DC. Needing a durable and reliable battery to run the **phonograph**, Edison significantly improved a French design in the late 1880s. The result was the so-called Edison–Lalande battery, which the **Edison Manufacturing Company** made and sold for decades for telegraph, telephone, and railroad signal systems, among other uses.

BATTERIES (STORAGE). Unlike **primary batteries**, storage batteries (also known as secondary batteries) generate electric current without depleting the electrodes or electrolytes. They can be recharged many times by reversing the current. In the early 1880s, Edison experimented with lead-acid storage batteries (invented in 1856 by Gaston Planté) for central stations to store electrical energy during the day for use at night, when demand was higher. He deemed them too inefficient for this purpose, although some electrical utility companies did adopt them to a limited extent.

When the advent of electric automobiles in the late 1890s created a new market for storage batteries, Edison decided to take a different approach to their design. Lead-acid cells were heavy and added significant weight to the vehicle, so Edison focused on developing a lighter battery that would also last longer. Drawing on his experience with the Edison–Lalande primary battery in the 1880s, he developed a storage cell with relatively light nickel and iron electrodes in an alkaline

SIXTY-SEVENTH YEAR

SCIENTIFIC AMERICAN

THE WEEKLY JOURNAL OF PRACTICAL INFORMATION

NEW YORK, JANUARY 14, 1911

THOMAS A. EDISON AND HIS IMPROVED STORAGE BATTERY

Edison posing with his storage battery, used to power this electric car (1911). *Scientific American 104 (14 January 1911): front cover.*

electrolyte. He organized the **Edison Storage Battery Company** in 1901 to manufacture it (starting in 1903). The battery sold well at first, but when it turned out to leak at its soldered seams and to lose its electrical capacity too quickly, Edison withdrew it from the market in late 1904. After spending several years and thousands of dollars on more experiments, he began producing an improved battery in 1909.

The new version was popular and performed well. Even so, it failed to expand the market for electric automobiles, which faced increased competition from gasoline vehicles—especially after **Henry Ford** introduced the mass-produced Model T. The market for electrics declined further during World War I as manufacturers focused on gasoline-power military use. Although Edison later made extensive experiments for Ford, his battery proved unsuited to the new electric starter systems for gasoline automobiles because it could not match the high initial charge of a lead-acid battery. Instead, Edison's alkaline storage batteries found extensive use for propulsion in boats and delivery trucks; for lighting in railway cars, lighthouses, miners' lamps, and isolated lighting plants; and for operating railroad signal systems. They even provided backup power for generating stations.

BELL, ALEXANDER GRAHAM (1847–1922).

Bell, an educator turned inventor, was born in Scotland just a few weeks after Edison's birth in Ohio. Their distinguished careers overlapped across two revolutionary inventions: the telephone and the **phonograph**. Bell's family lineage was steeped in the study and teaching of speech, and he was versed in the acoustic studies of **Hermann von Helmholtz** when he arrived in **Boston**'s fertile mix of craft, innovation, and entrepreneurship in 1871—two years after Edison left that city. In Boston, Bell embarked on creating acoustic telegraph systems. The project led to an association with attorney **Gardiner Hubbard**, his eventual marriage to Hubbard's daughter, and, in 1874–1875, the creation of an electric system for transmitting complex sound patterns—the telephone. Bell's instruments amazed visitors to the 1876 **Centennial Exhibition** in Philadelphia, but their signals—generated by **electromagnetic** induction from the physical force of sound waves—were too weak for the practical transmission of human speech (especially sibilant sounds like "sh" and "th"). Edison, inspired by the telephone's possibilities and its glaring weakness, devised his **carbon-button transmitter** as a more useful alternative. Then in 1878, Edison developed his **electromotograph receiver** in the hope of further getting around Bell's patents, but the instrument was only ever used in Great Britain. Western Union sold the Edison telephone patents to the Bell Company a year later, but the Edison transmitter (and others like it) dominated the industry for decades.

In 1880, the telephone earned Bell the French government's Volta Prize, worth about $10,000. Bell used the money to set up a laboratory in Washington, D.C. He hired Charles Sumner Tainter, a skilled Boston instrument maker, and together they worked on the phonophone—an instrument for transmitting sound via light (and a subject of some later interest to Edison). Bell also recruited Chichester Bell, a younger cousin with medical and scientific training. In 1881, the three turned their attention to sound recording, a field that Edison had abandoned. They focused on the recording medium (eventually replacing Edison's unsatisfactory tinfoil with a wax compound), the means of recording in the wax, and making duplicate recordings. "Alec" Bell moved on to other projects, but by 1886, Tainter and the younger Bell produced the **graphophone**—a sound-recording and playback machine that looked and worked (according to the eye) much like the phonograph. These similarities and the Volta trio's efforts to commercialize their instrument goaded Edison to resume development of the phonograph.

Bell helped restart the journal *Science* and published it (with Hubbard) after Edison withdrew his support; he also helped create the National Geographic Society. The **National Academy of Sciences** elected him a member in 1883. In later years, Bell turned his attention to heavier-than-air flight and to formulating eugenic principles from his experiences with animal husbandry and the families of his deaf students.

BERGMANN & COMPANY. This firm was one of four shops licensed to manufacture electric lighting equipment under Edison patents. It was the exclusive maker of sockets, fixtures, fuses, switches, meters, and similar articles for the Edison lighting business from 1881 to 1889, work that comprised about half its business. For several months in 1883, it also operated a Wiring Department to install interior wiring until the **Edison Construction Department** assumed those duties. In April 1881, **Bergmann & Company** succeeded S. Bergmann & Company (a partnership of **Sigmund Bergmann** and **Edward Johnson**), with Edison taking a one-third interest. Edison became president, but Bergmann ran the business himself. Expanding rapidly, the firm moved in 1882 from its factory on Wooster Street in New York City, where it employed about fifty

men, to a bigger one (acquired from a rival through a straw purchaser) on Avenue B. The new facility was promptly enlarged and soon hosted 350 workers. Edison also had a laboratory on the top floor from 1882 to 1886. Sigmund Bergmann left the company in early 1889 as it was being absorbed into the **Edison General Electric Company**.

BERGMANN, SIGMUND (1851–1927). A major manufacturing partner of Edison, Bergmann left his native Thuringia (later part of Germany) for the United States in 1869. Within a year, he was working as a skilled machinist for Edison in **Newark, New Jersey**. Bergmann opened his own shop in **New York City** in 1876, making telephones and **phonographs** for Edison, among other clients. **Edward Johnson** entered the business as a silent partner in 1879. Wishing to enlarge their capacity to make sockets, fixtures, instruments, and similar items for the Edison lighting system, Bergmann and Johnson took in Edison as a third partner in 1881 and reorganized as **Bergmann & Company**. Shrewd in business matters, Bergmann personally ran the thriving enterprise (occasionally clashing with Edison) until 1889, when he sold out to the new **Edison General Electric Company**. He then opened the Bergmann Electric and Gas Fixture Company in New York. Bergmann relocated to Berlin in 1890 and started yet another firm, the Bergmann Electrical Works, that became one of the largest of its kind in Germany. Early in the next century, he created the Deutsche Edison Akkumulatoren Company to manufacture and sell an alkaline storage (secondary) battery based on Edison's design. The battery was intended for electric cars and trucks, which Bergmann eventually produced himself.

BERLINER, EMILE (1851–1929). An immigrant from Hanover, Germany, with no formal scientific training, Berliner had a distinguished inventive career in the United States. He and Edison had little or no personal interaction, but Berliner made two major inventions that overlapped significantly with Edison's. The first was a telephone transmitter (a form of "microphone"), intended as an alternative to the transmitter of **Alexander Graham Bell**. Bell's instrument used the power of sound waves to generate electrical signals, but they were too weak to communicate intelligible speech. Berliner instead used sound waves to modulate the stronger currents produced by batteries. The key to this modulation was the loose contact between two electrodes, one of which received pressure from the transmitter's vibrating diaphragm. The changing pressure of one electrode against the other varied the resistance between them and thereby the strength of the signal through the circuit. Berliner filed a patent caveat for the device in April 1877 in which he mentioned the usefulness of bits of carbon for creating loose contact. He assigned his rights to the Bell Company (later the **American Bell Telephone Company**), for which he then worked for several years. Meanwhile, Edison was creating a variable-resistance telephone that became his **carbon-button transmitter**. He filed his first patent application in July 1877 and assigned his rights to **Western Union**. The question of whether Berliner or Edison "invented" the microphone vexed the Patent Office and dragged through the courts for years. The U.S. Supreme Court awarded legal credit to Edison in 1892—long after Western Union had ceded its telephone business to the Bell Company.

Of greater practical significance to Edison was Berliner's invention (in the late 1880s) of the **gramophone** and ancillary techniques for copying sound recordings. Edison initially dismissed the gramophone ("doesn't amount to anything") and disparaged Berliner's recording quality ("[he] never got the fine overtones") (Alfred Tate to Henry Villard, 10 December 1889, *TAED* LB035043; TAE to Patrick Delany, 22 December 1890, *TAED* LB046208). However, the gramophone and its flat discs (marketed after 1901 by the **Victor Talking Machine Company**) became formidable competitors, eventually overtaking the popularity of Edison's **phonograph** and its cylinder recordings. *See also* MICROPHONE CONTROVERSY.

BIRKINBINE, JOHN (1844–1915). A world-renowned engineer and mining expert,

Birkinbine worked for Edison as a consultant in iron mining and processing from July 1888 to July 1892. When Edison contracted with him, Birkinbine was consulting with the New York mining firm of Witherbees, Sherman & Company, which controlled mines in upstate New York. Birkinbine gave well-informed encouragement to Edison's ambitious—and ultimately unsuccessful—effort to revitalize and automate iron mining in the Northeast. He was also instrumental in promoting Edison's ore separators and wrote several articles highlighting the machines, including one coauthored with Edison though written largely by Birkinbine in 1889 ("The Concentration of Iron-Ore"). During his long career, he consulted or worked on the development of iron works, furnaces, mines, and hydraulic projects throughout North America. Birkinbine also worked for the U.S. Geological Survey and the Bureau of Mines and served as president of the Franklin Institute from 1897 to 1907. *See also* NEW JERSEY AND PENNSYLVANIA CONCENTRATING WORKS (NJPCW); ORE (IRON) MILLING.

BLISS, GEORGE HARRISON (1840–1900). Bliss was a Chicago-based promoter of Edison's inventions and business interests in the United States and Europe. His prolific correspondence illuminates Edison's affairs in the crucial years from 1876 to 1886, and his short biography of Edison, printed in the *Chicago Tribune* in April 1878 (and quickly revised and republished), was used extensively by later biographers and is the source of many errors and misconceptions. Formerly an electrical manufacturer and general agent of the **Western Electric Manufacturing Company**, Bliss became general manager of the **electric pen** business in 1876. In short order, he took on major duties promoting and selling Edison's duplicating ink, the **phonograph**, and the telephone. Appointed western agent of the **Edison Electric Light Company** at the end of 1881, he also set up a Chicago office of the **Edison Company for Isolated Lighting**. He helped form the Western Edison Company in 1882 to sell lighting systems in Illinois, Iowa, and Wisconsin, and he managed its affairs until 1886.

BLOCK, JULIUS HENRY (1858–1934). A British subject born in South Africa, Block was the proprietor of the Moscow-based J. Block Company, an importer of machinery and hardware that specialized in bringing new technologies to Russia. Among those technologies was Edison's wax-cylinder **phonograph** of 1888. Block—an accomplished amateur pianist—wished to acquire one to show to the tsar and record some of Moscow's top classical musicians (such as Sergey Taneyev, Anton Arensky, and Josef Hofmann). During a ten-day visit to the United States in 1889, he visited Edison's **laboratory** in **West Orange, New Jersey**. Edison personally instructed him on the use of the phonograph and saw to the careful packing of the machine, its battery, and selected recordings. Block returned to Russia and introduced the new machine first to Pyotr Ilych Tchaikovsky and then to Tsar Alexander III. He made some of the earliest recordings of Russian classical music and recorded the speaking voices of Anton Rubinstein, Leo Tolstoy, and Tchaikovsky. Block later moved to Berlin and made more early recordings of famous musicians. After World War I, he distributed these recorded cylinders among museums in Warsaw, Berlin, and Bern. The cylinders in Poland and Germany were thought to have been destroyed in World War II, but archivists at St. Petersburg's Institute of Russian Literature found them in the late twentieth century.

BÖHM, LUDWIG K. (1859–1917). Edison's advertisement in a German-language newspaper for a skilled glassblower was answered by Böhm, who joined the **laboratory** in **Menlo Park, New Jersey**, in August 1879. A native of Thuringia, Germany, Böhm studied under and worked with Heinrich Geissler, the renowned University of Bonn instrumentalist who created a mercury pump and electrical spark tube. At Menlo Park, he fabricated (and helped design) apparatus for Edison's incandescent lamp research, notably a hybrid mercury pump that produced the high vacuum critical to a successful lamp. He blew globes for Edison's earliest lamps and assisted with experiments on the vacuum preservation of perishable food. In August 1880, Edison assigned him to the new electric lamp factory. Apparently unhappy there and having quarreled with **Charles Batchelor**, Böhm quit in October and went to work for Edison's rival, **Hiram Maxim**. He later worked for the American Electric Light Company before returning to Germany in 1882 for one or more advanced degrees. Böhm returned to the United States in 1887, did a brief stint with the **Thomson-Houston Electric Company**, and became a consulting electrical and chemical expert in New York. He obtained at least fifteen U.S. patents. *See also* VACUUM TECHNOLOGY.

BOSTON. Thomas Edison arrived in Boston in March 1868 as a skilled press operator with inventive ambitions. By the time he left the city a year later, he had become a professional inventor. Edison's transition was made possible by his entry into one of the oldest and most sophisticated communities of telegraph manufacture and invention in the country. **Western Union** had four offices in the city, and the rival Franklin Telegraph Company had three. Edison soon discovered a host of telegraph and electrical manufacturers and inventors in Boston and reported on their work in articles for *The Telegrapher*. Boston manufacturers like Charles Williams Jr., who ran one of the city's largest telegraph shops, provided Edison and other inventors a place to experiment and extended credit for work and materials. The bustling commercial center also had numerous local entrepreneurs and company officials willing to provide direct financial support.

Inspired by the technologies on display in the city's telegraph shops and in practical use, Edison experimented with nearly every form of telegraph and electrical apparatus during his year in Boston. These included a double transmitter, a self-adjusting relay, a method of **automatic telegraphy**, a **stock printer**, a dial telegraph, a fire alarm telegraph, a facsimile telegraph, and an electric **vote recorder**. While the vote recorder was the subject of Edison's first patent, it was his second patent for a stock printer that marked his first successful invention. Local entrepreneurs funded a stock-quotation service using Edison's printer that began operation in January 1869. The banking and brokerage house of Kidder, Peabody and Company was the first of twenty-five subscribers to the new service. In addition, Edison used his printer and (briefly) his dial telegraph to provide rapid communication for several Boston businesses on private lines between their offices, factories, and warehouses. These enterprises struggled due to inadequate funding for improving the technology. In April 1869, Edison moved to **New York City**—the nation's largest city and the center of its booming telegraph industry—to make experiments on his double transmitter.

BROWN, HAROLD PITNEY (1857–1944). An electrician and inventor, Brown campaigned strenuously against **alternating current** (AC) and was, for a time, allied with Edison and the **Edison Electric Light Company** in the **War of the Currents**. Brown gained experience with AC at Chicago power plants from 1879 to 1884, and he later formed a company specializing in long-distance transmission. A spate of ghastly accidents sensitized him to the dangers of high-tension AC lines for arc lighting. As lighting companies began adapting high-voltage AC for incandescent lighting in the late 1880s, Brown began a provocative campaign against power lines carrying more than 300 volts in populated areas. In one instance, he published a letter in the *New York Evening Post* in June 1888 warning that AC lighting systems posed a "constant danger from sudden death" (*TAEB* 9:208). That same month, he contacted Edison after reading about **electrocution (animal experiments)** at the **laboratory** in **West Orange, New Jersey**. At Edison's invitation, Brown joined **Arthur Kennelly** in further experiments in 1888 and 1889, and he later published an account of this research. Brown's leading role in particularly gruesome animal demonstrations in New York City drew sharp press criticism and made him the target of a countercampaign by the **Westinghouse Electric Company**. However, his seemingly independent opposition to AC proved useful to Edison and Edison Electric, and Edison praised him as a "thoroughly competent Electrical Engineer" whose reports were "accurate and truthful" (*TAEB* 9:561). But in August 1889, the *New York Sun* published forty-seven letters (or excerpts) presumably stolen from Brown's desk. The correspondence was from, to, or about Brown and his connections with various electric lighting companies, and it showed his symbiotic relationship with Edison Electric. However, the letters did not prove the claim by the *Sun* that Edison (against his sworn testimony in the death penalty case of William Kemmler) had paid Brown for his services.

BURROUGHS, JOHN (1837–1921). A member with Edison of the **Vagabonds** group of famous men who made stylized annual retreats into nature, Burroughs was an esteemed naturalist and author deeply influenced by the Transcendentalist writings of Ralph Waldo Emerson. He met **Mina Miller Edison** in 1906 at a party hosted by painter Orlando Rouland (who later did a portrait of Edison), and the two formed a friendship based on their mutual interest in birds. However, by that time, Burroughs had shifted his prolific literary output to matters of philosophy and religion informed by recent science, topics presumably of more interest to Edison than ornithology. Edison and Mina visited Burroughs at his home in the Catskills in 1913 and repaid his hospitality the next year when Burroughs and **Henry Ford** came to Florida. The latter gathering marked the informal start of the Vagabonds' annual peregrinations. Burroughs was also known for his close friendships with Walt Whitman and, in his last two decades, with Theodore Roosevelt.

BUSINESS PHONOGRAPH. Almost as soon as he invented the **phonograph** in 1877, Edison began trying to develop a model for business dictation. He tried a flat disc covered with tinfoil, but it did not work. In 1888, in conjunction with the overall redesign of the phonograph, he brought out a wax-cylinder dictation model run by electricity, but it proved too complex for ordinary business use. Edison introduced a more reliable and easy-to-use dictating machine in 1905, putting it under the management of Nelson Durand, who would lead the subsequent development of the Edison business phonograph. Between

1911 and 1914, Durand worked with Newman Holland of the staff at the **laboratory in West Orange, New Jersey,** to develop a sturdier and more accessible business phonograph that incorporated an attachment called the Transophone, intended to make repeating and correcting easier. Holland also helped develop a pet project of Edison's known as the Telescribe. The Telescribe was a combination telephone and phonograph for recording both sides of a telephone conversation, but it never worked well. The business phonograph, renamed the Ediphone in 1918, developed into an integrated office system with separate machines for dictating, transcribing, and shaving cylinders.

C

CALAHAN, EDWARD AUGUSTIN (1838–1912). A **Boston**-born telegraph inventor, Calahan worked as an operator and manager first for the American Telegraph Company and then for **Western Union Telegraph Company**. He invented the first **stock ticker** while working as a draftsman and assistant to Western Union engineer **Marshall Lefferts**. This small printing telegraph became the mainstay of the **Gold and Stock Telegraph Company**'s inventory of printers until Edison's Universal stock printer superseded it in the early 1870s. In 1871, Calahan invented the district telegraph system, which enabled users to signal messenger, police, fire, and other services; he organized the **American District Telegraph Company**, forerunner of the **American District Telegraph Company**, to promote the system. For a brief period in the mid-1870s, Edison had a competing district telegraphy system operated by the Domestic Telegraph Company.

CARBON. *See* CARBON-BUTTON TELEPHONE TRANSMITTER; ELECTRIC LAMP; TASIMETER.

CARBON-BUTTON TELEPHONE TRANSMITTER. After hearing accounts of **Alexander Graham Bell**'s telephone demonstrations at the 1876 **Centennial Exhibition** in Philadelphia, Edison began experimenting with the new technology. He soon identified the transmitter as a weak point. Bell's device had a permanent magnet that, when vibrated by sound waves, induced a variable current ("undulatory," according to Bell) in an electromagnet.

The current went out to the line and, at the other end, underwent a reverse process in the receiver to reproduce sound. The transmitter could generate a current only as strong as the force of air on the transmitter diaphragm, severely limiting the distance over which it could be used. In contrast, Edison used an external—and stronger—energy source: a battery. He applied battery current to the line and used carbon as a resistance medium to vary the current's strength in response to pressure produced by sound waves. In doing so, Edison drew on his earlier experiments with powdered carbon in glass tubes to simulate the resistance of submarine telegraph cables. Those artificial cables proved unreliable because noise and movement affected the resistance of the carbon, but that kind of sensitive variable resistance was just what Edison needed for the telephone. By the end of 1877, he had devised a transmitter with a small button of lampblack carbon beneath the diaphragm. Sound waves moved the diaphragm, varying pressure on the button and thereby its resistance to the battery current on the line. He finalized details of the diaphragm-and-button arrangement early the next year.

Edison's assistants made the carbon for his early telephones in a small shed at the **laboratory** in **Menlo Park, New Jersey**. They burned kerosene lamps continuously, scraping the accumulated lampblack from the glass and pressing it into buttons. **Western Union** marketed Edison's transmitter until it decided to leave the telephone business. In 1885, Edison developed an improved carbon transmitter for the **American Bell Telephone Company** that used granules of roasted anthracite coal instead of lampblack. Carbon transmitters remained in wide use until the advent of digital telephones in the 1980s.

CEMENT HOUSE. In 1906, as his Edison Portland Cement works in Stewartsville, New Jersey, began to reach full operating capacity, Edison conceived the idea of producing moderately priced working-class housing using cement. He worked out and patented a plan to form houses in cast-iron molds that could be quickly assembled and taken apart. Edison did not plan to commercialize this system himself but instead freely licensed the technology to others. The American Building Corporation built the first home using a system based on Edison's patents in 1909—in Montclair, New Jersey, where corporation president Frank D. Lambie lived. Lambie subsequently patented an improved steel mold system he used to build cement homes in Pennsylvania. In the late 1910s and early 1920s, Charles Ingersoll licensed the system from Edison but used wooden molds to reduce the price of the homes he built in Union and Phillipsburg, New Jersey.

CENTENNIAL EXHIBITION. The Centennial Exhibition in Philadelphia, commemorating 100 years of American independence, was the first world's fair staged outside of Europe. It opened on 10 May 1876 to much fanfare, including a speech by President Ulysses S. Grant and a "Centennial Exhibition March" composed by Richard Wagner. Over the next six months, roughly 10,000,000 people attended the Exhibition, which showcased technological inventions and innovations such as **Alexander Graham Bell**'s telephone, the **typewriter**, and the 200-ton Corliss steam engine. While Edison did not attend, the **Atlantic and Pacific Telegraph Company** used his **automatic telegraph** system to run the exhibition's public telegraph office. A committee headed by British physicist **Sir William Thomson** awarded a prize medal to Edison's system, praising it as "a very important step in land-telegraphy" (*TAEB* 3:57). Thomson also

lauded Edison's **electric pen** and autographic press as an invention of "exquisite ingenuity and . . . usefulness" (*TAEB* 3:159 n. 10). Another inventor drawing plaudits at the Exhibition was Bell, whose demonstration of the telephone not only stunned Brazilian emperor Pedro II—"My God; it talks!" he allegedly cried as he dropped the device—but also inspired Edison to intensify his research on telephony (Picker 2003, 101). By the middle of 1877, the telephone was the principal focus of work in the **laboratory** in **Menlo Park, New Jersey**.

CHAUTAUQUA INSTITUTION. The name "Chautauqua" designates both a physical place in western New York State and the institution based there. **Lewis Miller**, a devout Methodist (and Edison's future father-in-law), and John Heyl Vincent, a Methodist bishop, founded the Chautauqua Institution in 1874 as an outgrowth of a successful experiment in out-of-school vacation learning. The Institution provided recreation, music, physical activities, and religious instruction to those with the means and inclination for uplifting summer leisure. It also sponsored nonresident programs, notably the Chautauqua Literary and Scientific Circle, a nationwide program of guided reading established in 1878 as an economical alternative to collegiate study. Immediately popular, the Institution grew to occupy more than 130 acres along the shore of Lake Chautauqua, where it hosted hundreds of attendees at its assemblies. Chautauqua held an important place in the collective life of the Miller family, who had a cottage there. Edison visited while courting the young Mina Miller in 1885, and the couple returned on numerous occasions thereafter. *See also* EDISON, MINA MILLER.

CHEMISTRY. Edison is usually known as an electrical inventor, but much of his experimental work—especially in telegraphy, electric lighting, batteries, and recorded sound—involved chemistry. According to his official biography, he "wondered how it was that he did not become an analytical chemist instead of concentrating on electricity" (Dyer and Martin 1910, 28), and retired chemist Byron Vanderbilt explored this aspect of Edison's career in his 1971 book *Thomas Edison, Chemist.* Edison acquired most of his chemical knowledge by himself, first in a makeshift lab in a baggage car of the Grand Trunk Railway while working on the train and then from his home (where he moved his equipment). Edison made experiments guided by a translated version of what his biographers identified as Carl "Fresenius's *Qualitative Analysis*" (Dyer and Martin 1910, 37), most likely either *System of Instruction in Qualitative Chemical Analysis* or *Elementary Instruction in Chemical Analysis.*

Edison's chemical experiments in the late 1860s and early 1870s focused on recording systems (for electrical **vote recorders**, facsimile telegraphs, and automatic telegraphs) in which an electrical discharge through a metal stylus would produce a mark by discoloring chemically treated paper. These experiments led Edison to begin a more thorough study of the chemical literature and to seek out Robert Spice, professor of chemistry and natural philosophy at Brooklyn High School, for private instruction in 1874.

Edison read the published chemical literature to expand his knowledge and to make his experiments more systematic. Henry Watts's *Dictionary of Chemistry and the Allied Branches of Other Sciences* became a standard reference about chemicals, minerals, and metals in Edison's **laboratories** and was the starting point for extended research on materials for lamp filaments, phonograph record materials, electrical insulation, and batteries. Edison turned to Watts and other publications to find the most promising chemicals or materials for the project at hand. Characteristically seeking to confirm existing knowledge by direct experiment, he also tried materials that published authorities suggested would not work well, especially in new physical conditions, such as electric lighting. Another important instance of Edison turning to the large body of published chemical knowledge came around the time of World War I. Faced with wartime embargoes of the carbolic acid, benzol, aniline, and other chemicals from Europe that he needed to make phonograph records, he read the literature to determine the best methods for making the chemicals himself. He then worked with his staff to design manufacturing processes and facilities.

Beginning with his electric light research, Edison hired PhD chemists, especially those educated at German universities, for his **laboratories** in **Menlo Park, New Jersey**, and **West Orange, New Jersey**. However, Edison sometimes complained that trained chemists were insufficiently practical, and he employed men with less formal education. Among them was **Reginald Fessenden**, a self-taught electrical engineer whom Edison put in charge of the West Orange chemical lab. Another was **Jonas Aylsworth**, who left college after one year and played a key role in developing phonograph record materials in West Orange. Aylsworth's work on phenol resins for disc records in the 1910s was at the cutting edge of materials research and polymer chemistry. However, a decade later, Edison had no chemists on his staff familiar with current work in this field. By then, he seems not to have appreciated the most recent chemical discoveries, and his final project—to find a domestic source of **rubber**—focused on natural materials rather than developing synthetic rubber.

CIVIL WAR TELEGRAPHY. The American Civil War was a significant event in Edison's early career. The war began while Edison was selling newspapers and candy to passengers on the Grand Trunk Railway trains between his home in **Port Huron, Michigan**, and Detroit. News of battles (in which thousands of Michigan volunteers fought) increased his newspaper sales. Edison took special advantage of this local interest during the Battle of Shiloh in April 1862, which included four Michigan regiments and produced the largest number of casualties to that point in the war. Edison arranged with one of the railroad operators to telegraph news of the battle ahead of his train. He then persuaded the editor of the *Detroit Free Press* to take a chance and give him papers to sell. Edison's plan worked even better than expected. Large crowds greeted the train at each station, enabling him to raise the price and still sell so many papers that he made "what to me was an immense sum of money" (Israel 1998, 16). Edison later claimed that this episode led him to learn telegraphy.

Edison had his first job as a part-time telegrapher in a local jewelry store in Port Huron by the end of 1862, and within six months, he was working as a telegrapher for the Grand Trunk Railway at Stratford Junction, Ontario. During the years 1863–1868, Edison became one of many young men who joined the ranks of itinerant or "tramp" telegraphers and drifted from city to city. While tramping was an economic necessity for many workers, the primary motive for those with highly valued skills, such as telegraph operators, was wanderlust rather than hardship. This was especially true during and immediately after the Civil War due to the high demand for skilled operators on commercial lines to replace those who had gone into the U.S. Military Telegraph Corps.

Edison initially found employment as a railroad telegrapher in Adrian, Michigan, and then in Fort Wayne, Indiana. By late 1864, he had honed his skills sufficiently to work for **Western Union**'s main Indianapolis office. In the spring of 1865, he moved to the Western Union office in Cincinnati and by September was promoted to operator first class. Later that fall, he became the regular press-wire operator in Memphis, Tennessee, for the South-Western Telegraph Company. Edison found that the end of the war had brought widespread lawlessness, gambling, and violence to the city, and the following spring, he moved to the Western Union office in Louisville, Kentucky, which had emerged from the war as a prosperous industrial center.

Edison and two companions decided in late July 1866 to seek adventure and fortune in Brazil, a country actively seeking American immigrants—especially telegraphers. Their journey ended in New Orleans, where a bloody race riot a few days earlier led to martial law and the commandeering of their Brazil-bound steamship. After a short visit to Port Huron, Edison returned to his position at the Western Union office in Louisville and then, briefly, back to the Western Union office in Cincinnati in the summer of 1867. He then moved to **Boston**

and worked in the main Western Union office for several months before resigning to begin his career as a professional inventor.

Like other ambitious telegraph operators, Edison moved through the ranks to join the elite press operating corps. Press operators were in a good position to gain insights into other fields, especially journalism and politics, as well as to acquire knowledge of business practice, which they could use to establish successful careers outside of telegraphy. The experience served Edison well as an inventor.

Throughout his years as an itinerant operator, Edison demonstrated interest in and aptitude for experimenting with and improving telegraph technology. His fellow operators described him as studious, someone who spent his free money and time on books and experiments. Edison focused especially on the self-adjusting repeaters and relays used to switch the weak incoming signal to a new transmitting circuit with its own battery so it could be forwarded to the next station. The longer transmission distances created by Western Union's consolidation of the nation's telegraph lines after the Civil War made improvements in these devices particularly important. Edison also experimented with secret signaling for military telegraphs and sending two messages on a single wire (**duplex telegraph**), thus increasing the capacity of capital-intensive telegraph lines. In Boston, he would expand his experiments to encompass fire-alarm and other local telegraph services.

CLARKE, CHARLES LORENZO (1853–1941).

An experienced draftsman and skilled engineer, Clarke helped design Edison's central station **electric light system** and the **generators** for large central stations and smaller isolated plants. A Bowdoin College classmate of **Francis Upton**, Clarke was among Edison's few college-educated assistants when he joined the staff of the **laboratory in Menlo Park, New Jersey**, in January 1880. In his first year, he oversaw construction and testing of the massive "Jumbo" direct-connected steam dynamo for lighting large cities. When Edison left Menlo Park for **New York City** in 1881, Clarke became chief engineer of the **Edison**

Electric Light Company, a position he held until early 1884. After stints in business and as a consulting engineer and patent expert, Clarke worked with the Board of Patent Control, a patent pool of the **Westinghouse Electric Company** and the **General Electric Company**. He spent the last two decades of his career at General Electric.

CLEMENS, SAMUEL LANGHORNE (1835–1910). Known to his reading public as Mark Twain, Clemens was a novelist, essayist, and entrepreneur celebrated by the late 1880s for creating *The Adventures of Tom Sawyer* and *The Adventures of Huckleberry Finn*, among other books. In May 1888, he asked to meet with Edison in the hope of obtaining a new wax-cylinder **phonograph** for dictating his novel, *A Connecticut Yankee in King Arthur's Court*. Clemens suffered from rheumatism in his shoulder, and he hoped the phonograph could spare him the discomfort of writing. Clemens visited the **laboratory** in **West Orange, New Jersey**, in June 1888 but left without a phonograph. Edison hoped to provide one by September, when the **Edison Phonograph Works** was to start making them, but by then, Clemens had canceled his order. Edison recalled Twain's visit in a 1927 letter to Cyril Clemens, president of the Mark Twain Society. Clemens told funny stories, he remembered, which were recorded on a phonograph. Edison later claimed these recordings were lost in the 1914 fire at his phonograph factory.

Although Clemens did not use a phonograph in writing *A Connecticut Yankee*, he did include it among the American inventions the novel's protagonist introduces into Arthurian England. Clemens himself finally secured a rented phonograph in 1891 to dictate *The American Claimant*, in which the machine again appears. The protagonist, a madcap scientist, devises the machine for maritime service so that it could play recordings of swearing, effectively relieving sailors of that duty. Although Clemens disliked having the phonograph in his creative process (calling it "just matter-of-fact, compressive, unornamental, and as grave and unsmiling as the devil"),

in his later years, he alternated dictation with longhand writing to ease his physical discomfort (Twain 1917, 2:543–44).

COLUMBIA. The steamship *Columbia* received a complete Edison electric lighting plant, the first outside the **laboratory** in **Menlo Park, New Jersey**. Built for **Henry Villard**'s Oregon Railway and Navigation Company for service between Portland, Oregon, and San Francisco, the *Columbia* was fitted out in **New York City** in April and May 1880. Edison's assistants supervised **Bergmann & Company**'s installation of four **generators**, fixtures, and wiring. Edison, his wife **Mary Stilwell Edison**, and several assistants attended a reception aboard the ship as the work neared completion. The system worked well save for the breakage of lamps, which Edison could not readily resupply to the Pacific coast. The ship and its lighting were featured in an illustrated article in *Scientific American* ("The Columbia," 22 May 1880, 42:326).

COMPAGNIE CONTINENTALE EDISON. This holding company was created in 1882 as the keystone of a complex organization to develop Edison electric lighting in much of continental Europe. European investors backed the Paris-based company. It licensed Edison's patents from the Edison Electric Light Company of Europe, a patent-holding entity in **New York City**. It then sublicensed the patents to local utilities and two sibling firms: the Société Industrielle et Commerciale Edison, which made lamps and other equipment near Paris, and the Société Électrique Edison, which installed isolated lighting. The company had some successes, notably in France and Italy, but generated disappointing revenues for the New York company and failed to build a Paris central station. Its limited revenues and the unwieldy administrative arrangements in Europe led **Henry Villard** to push the company to absorb the manufacturing and isolated firms in 1886. The combined firm, facing competition from Edison's own manufacturing shops in the United States and **alternating current** lighting systems in Europe, acquired a limited ability to make and sell non-Edison equipment. It did eventually build and operate a Paris central station.

CONSTABLE, JOHN PIERREPONT (1888–1926).

Born in Utica, New York, Constable attended the Massachusetts Institute of Technology (MIT), where he was a classmate of **Charles Edison**. After graduating from MIT in 1914, he worked at the Engineering Department of Edison's **laboratory** in **West Orange, New Jersey**. His job involved designing the Amberola **phonograph** and planning for its mass production. Constable became assistant chief engineer in 1915, taking the title of chief engineer in 1916, effectively replacing **Miller Reese Hutchison**. During World War I, he served as captain of the Laboratory Company of the Edison Battalion. Constable was elected a trustee of the Thomas A. Edison Association in 1918. Edison fired him at the start of 1921 after **William Meadowcroft** found that he was involved in making unauthorized copies of Edison's internal employee questionnaire.

COSTER, CHARLES HENRY (1852–1900).

A financier important to Edison's electric lighting enterprise and to American business in general, Coster began his career in the counting room of the **New York City** import merchant house Aymar & Company. When the shipping company Fabbri & Chauncey acquired the firm in 1872, Coster attracted the favorable notice of its principal, Egisto Fabbri, who was a partner in **Drexel, Morgan & Company** and deeply involved in Edison's lighting projects. Following Fabbri's financial failure in 1883, Coster replaced him as a Drexel Morgan partner and trustee in the **Edison Electric Light Company** and **Edison Company for Isolated Lighting**. Coster helped to streamline the Edison companies in 1884 amid a general financial crisis and Edison's decision to wrap up the **Edison Construction Department**. The reorganization also gave the Edison Electric Light Company and its investors equity interests in Edison's now-profitable manufacturing shops, a fact that later contributed to the formation of the **Edison General Electric Company** (EGE). Coster played a key role in forming the **General Electric Company** in 1892 through the merger of EGE and **Thomson-Houston Electric**. By that time, he had helped Drexel Morgan reorganize and consolidate numerous rail companies (a process called "morganization") and was renowned as "[**J. P.**] **Morgan**'s right arm" (*TAEB* 7:586 n. 7).

CRAIG, DANIEL HUTCHINS (1811–1895).

A former head of the Associated Press, Craig became the leading American proponent of **automatic telegraphy** for transmitting press reports after he obtained a financial interest in George Little's automatic telegraph system in 1869. The following year, Craig approached Edison, who began to develop his own improved automatic system. To support Edison's work, Craig brought in former assistant secretary of the treasury **George Harrington** and railroad executive and financier William J. Palmer and his associates, who joined to form the **Automatic Telegraph Company**.

CROOKES, SIR WILLIAM (1832–1919).

This English physicist and chemist had limited direct contact with Edison, but his accomplishments as a skilled experimenter significantly influenced Edison's work on the electric light. Edison began reading his work in the mid-1870s and, over the years, inserted into his scrapbooks published accounts of Crookes's investigations of high-vacuum phenomena. Of particular interest were Crookes's creation of a "light mill" (the radiometer) in 1875 and his postulation of a fourth state of matter (termed "radiant matter") manifested by electrical discharges in rarefied gases. The radiometer proved to be a popular instrument, and Edison suggested manufacturing it at his lamp factory, but the idea went nowhere. Of greater significance, Edison and glassblower **Ludwig Böhm** drew on techniques and apparatus created by Crookes and his assistant Charles Gimingham to design their own mercury pumps for the high vacuum levels crucial to the incandescent lamp. When, in 1880, Edison noticed strange effects inside the lamp globes, including an eerie glow around the clamps holding the filament to lead-in wires and a gradual darkening of the glass, he turned to the idea of radiant matter. Crookes had theorized radiant matter as streams of negatively charged molecules propelled away from an electrified wire. In the mid-1880s, Edison himself created "radiant matter"—what would later be called plasma—in his attempts to make the **direct conversion** of coal into electricity. Crookes also contributed to the intellectual milieu of the 1870s and 1880s by publishing *Chemical News*, an authoritative journal that counted Edison among its subscribers. Edison used its pages in 1878 to press his priority claims in the **microphone controversy**, but Crookes did not become personally drawn into the dispute. *See also* EDISON EFFECT; VACUUM TECHNOLOGY.

CUTTING, ROBERT LIVINGSTON, JR. (1836–1894). A wealthy **New York City** financier, Cutting was an investor in and director of numerous Edison companies. After graduating from Columbia College, he joined the brokerage firm of his father (Robert L. Cutting & Company), who also invested in Edison telephone and electric light enterprises. The younger Cutting later became a partner in a different firm. He was an incorporator of the **Edison Electric Light Company**, the **Edison Electric Illuminating Company of New York**, and the Edison Telephone Company of Europe, and a founding director of the **Edison Phonograph Works**. He became particularly involved with Edison's mining projects as an original investor in the **Edison Ore Milling Company** and the **New Jersey and Pennsylvania Concentrating Works**. It was Cutting who brought the French actress Sarah Bernhardt to meet Edison at **Menlo Park, New Jersey**. A prominent Democrat, he served on the Committee of Seventy (1870), which investigated corruption in New York City's Tammany Hall political machine.

D

DANIELS, JOSEPHUS (1862–1948). As secretary of the navy for Woodrow Wilson, Daniels oversaw creation of the **Naval Consulting Board** in 1915 (under prompting by **Miller Reese Hutchison**) to prepare the nation for war, naming Edison as its head. Daniels also championed Edison's S-type **storage battery** in submarines as replacements for lead-acid cells that risked creating chlorine gas if the acid electrolyte mixed with seawater. However, Edison's batteries had their own hazards from hydrogen gas and played a role in a fatal explosion aboard a submarine in the Brooklyn Navy Yard in January 1916. Partly as a result, Daniels could not persuade the navy's Bureau of Steam Engineering to adopt the Edison battery, though he did shield Edison from direct blame for the disaster. Daniels entered public life as an editor and publisher of several newspapers in his native North Carolina, in which roles he aided the violent overthrow of the elected biracial government of Wilmington, North Carolina, by a white mob in 1898. Daniels later served in the administration of Franklin Roosevelt, his one-time assistant navy secretary, as U.S. ambassador to Mexico (1933–1941).

DEAFNESS. Edison apparently first became aware of his hearing loss as a boy while selling newspapers and candy on the Grand Trunk Railway. He would later tell two stories about its origin. One ascribed it to a conductor who angrily boxed his ears after his chemistry set caused a small fire aboard the train. In the other version, a conductor lifted him by the ears aboard a moving train, and Edison felt a pop. Biographers have also ascribed his hearing loss to a presumed case of scarlet fever. More recently, a medical expert has attributed it to fenestral otosclerosis, a common cause of the kind of progressive conductive hearing loss that Edison experienced. (He did not become almost totally deaf until the mid-1920s.) Over the course of his life, Edison found that his poor hearing reduced distractions and allowed him to concentrate better. It gave him a convenient excuse to skip meetings and social events. It also presented hindrances. Edison loved going to the **theater** but found it hard to hear unless seated in the front row. Family members and colleagues increasingly found themselves hollering into his ear. Partly for himself and partly for the numerous hard-of-hearing persons who importuned him for help, Edison periodically experimented with hearing aids and even tried commercial designs, but none was satisfactory.

Edison's deafness did not prevent him from working on acoustic technologies, such as the **carbon-button telephone** and the **phonograph**. He thought it even conferred some advantages. He attributed his 1870s telephone improvements to the necessity of making the instrument loud and clear enough for him, though he sometimes had to rely on the judgment of his assistants regarding the "finer articulations" (*TAEB* 3:518). During the 1910s, he personally conducted trials of every singer and musician who recorded for his company by biting into the wood of the piano or phonograph to conduct sound vibrations to his inner ear. He asserted that doing so allowed him to hear better than those with normal auditory perception because his inner ear was more sensitive, as it "has been protected from the millions of noises that dim the hearing of ears that hear everything" (Israel 1998, 435). Edison's odd way of listening to music likely influenced his tastes. Certain notes and instruments hurt his ears, which may account for his dislike of singers' vibrato.

DELMONICO'S. One of Edison's favorite places to eat, Delmonico's was among the earliest restaurants in the United States. It opened on William Street in Manhattan's Battery section in 1830 or 1831 as an adjunct to a pastry shop established by Swiss immigrants John and Peter Delmonico. Considered **New York City**'s finest restaurant for much of the nineteenth century, it attracted an upscale clientele. By Edison's time, it had branched out to four locations, one of which, at Fifth Avenue and 26th Street, was frequented by the inventor and his associates. This six-story restaurant operated under the management of renowned chef Charles Ranhofer.

Delmonico's figured in several important occasions in Edison's life. When he invited dozens of New York City officials to his **laboratory** at **Menlo Park, New Jersey**, in December 1880 to witness his new electric light system, Delmonico's catered the event. In February 1886, several close associates, including **Charles Batchelor**, **Edward Johnson**, and **Samuel Insull**, threw Edison a bachelor party at the restaurant prior to his marriage to **Mina Miller**. When illness confined Edison to his **Llewellyn Park** home a year later, Insull arranged for the headwaiter to bring his favorite dishes. Insull himself, who also enjoyed Demonico's, received a farewell celebration there when he left the Edison world and moved to Chicago in June 1892. The last Delmonico-owned restaurant (on Fifth Avenue at 44th Street) closed in 1923. The current (2022) Delmonico's opened in 1998.

DEUTSCHE EDISON GESELLSCHAFT (DEG). Formally the Deutsche Edison Gesellschaft für angewandte Elektricität (the German Company for Applied Electricity), this Berlin-based company was created in 1883 to manufacture, sell, and install electric lighting equipment in Germany under Edison's patents. Inspired by the Edison exhibit at the **Paris International Electrical Exposition (1881)**, electrical entrepreneur Emile Rathenau acquired the German patent rights from Edison's European companies in Paris. He and Oskar von Miller, a recent engineering graduate, also animated by the 1881 exhibition, organized DEG in March 1883. DEG was to use Edison's incandescent lighting but was free to employ arc lighting systems by other inventors. It licensed the manufacture of Edison lamps to Siemens & Halske, which also supplied other electrical equipment. With **William Hammer** as its chief engineer, DEG built a small demonstration electrical plant in Berlin followed by a larger central station. When DEG's success and broad ambitions created conflicts with the sleepier pan-European enterprises in Paris, **Henry Villard** orchestrated a grand restructuring of the Edison lighting interests in Europe. One consequence was the reorganization of DEG as the Allgemeine Elektrizitäts-Gesellschaft (AEG, or the General Electric Co.) in May 1887. The removal of Edison's name reflected the reality that AEG was neither restricted to using his patents nor beholden to his European companies. *See also* COMPAGNIE CONTINENTALE EDISON.

DICK, ALBERT BLAKE (1856–1934). Dick founded the A. B. Dick Company as a Chicago lumber wholesaler in 1884. To reduce time spent on office work, Dick devised a method of copying handwritten documents by creating a perforated stencil on waxed paper and squeezing ink through the holes. He tried to secure a U.S. patent only to find that Edison had already gotten one (no. 224,665) in 1880 in conjunction with the **electric pen**. In 1887 Dick reached a patent-licensing agreement with Edison and began to market the copying technology as the Edison Mimeograph. He soon sold off the lumber business and concentrated on improving, manufacturing, and marketing the mimeograph and other office machines—an enduring and profitable enterprise. Meanwhile, Edison and Dick collaborated to develop processes and materials (including inks) for mimeographing typed documents. Edison also sent Dick to Europe in spring 1889 to investigate the market and suppliers for his **talking doll**, and a few months later Dick joined Edison's entourage at the **Paris Universal Exposition (1889)**. *See also* TYPEWRITER.

DICKSON, WILLIAM KENNEDY LAURIE (1860–1935).

Dickson (often "Dixon" to Edison) played important roles in two major Edison projects: **ore processing** and **motion pictures**. He is often regarded as a co-inventor of the latter. Born in France of Scotch-English ancestry (his mother may have been American by birth), Dickson asked Edison for a job from London in 1879 and again from **New York City** four years later. By then he was an educated and well-traveled young man, and Edison assigned him to the Testing Room of the **Edison Machine Works**. Dickson was running the Testing Room within a year. He then followed Edison to his successive **laboratories** in New York City; to the lamp factory near **Newark, New Jersey**; and to **West Orange, New Jersey**. At West Orange, Dickson took charge of experiments on processes for separating iron and gold from their ores. His expertise with iron ore was such that in 1890 Edison sent him to troubleshoot the new plant of the **New Jersey and Pennsylvania Concentrating Works**, giving him authority over even the nominal superintendent there. Dickson designed the belt separator used in the "refining mill"—the final stage of concentration—for which he became one of the few assistants to take out patents jointly with Edison.

Dickson also had photographic skills (his father had been a portrait artist), and Edison designated him as the laboratory's photographer. Familiarity with the equipment and processes made him a natural collaborator when Edison turned his attention to motion pictures in late 1888. (Dickson sometimes mis-remembered or mis-reported dates when it suited him, and he attributed the start of this work to 1887.) His first assignments involved microphotography—tiny images on the out-

side surface of a rotating drum. During the next summer, Edison and Dickson experimented with narrow strips of celluloid film as a substitute for the rotating drum. They may have had a prototype strip-film **motion picture camera** (kinetograph) when Edison left in early August for the **Paris Universal Exposition (1889)**, and Dickson probably had a functioning kinetoscope to demonstrate on Edison's return in October. Edison included the new camera and viewing apparatus—both using sprocket wheels to engage perforated film—in a patent caveat a few weeks later.

Over the next several years, Dickson made significant contributions to the development of a practical kinetograph and kinetoscope. In 1894–1895, he and Edison created the **kinetophone** to synchronize moving images with recorded sound. Dickson left Edison's employ in 1895. The next year, he and several partners founded the **American Mutoscope and Biograph Company**, which became a major competitor to Edison in the production and distribution of motion pictures. Dickson returned to the United Kingdom in 1897 and embarked on a new career as a pioneering film director for the Biograph company.

In 1894, Dickson and his sister Antonia Dickson published a book-length review of Edison's life and career and a magazine article, "Kineto-Phonograph" (the kinetophone). The next year, they produced a shorter *History of the Kinetograph, Kinetoscope, & Kinetophonograph*. Their accounts are useful historical sources despite some dubious claims and chronological inaccuracies.

DIRECT CONVERSION.

Edison used this term for a hypothetical process of producing electricity from coal without mechanical intermediaries, such as steam engines and dynamos. The idea was to capture the estimated 90 percent of energy wasted by conventional generation methods. Edison took up the subject (also called "direct oxidation") in early 1882, when he designed something like a high-temperature battery with an oxidizing carbon electrode. Returning episodically to direct conversion in 1883 and 1884, he developed versions of what came to be known decades later as **fuel cells**. One experimental process reportedly produced a "very strong current" in the summer of 1884 (*TAEB* 7:253). Edison hoped to show it at the **Philadelphia International Electrical Exhibition (1884)** in the fall but changed his mind after an explosion blew out his **laboratory** windows. Frustrated but still optimistic, he looked forward to a breakthrough that would "revolutionize the industrial world" (*TAEB* 8:18 n. 3). Related experiments in 1887 led him to develop instead a **pyromagnetic generator and motor**. He spoke in 1893 of making direct conversion his next big project after iron-ore milling. He did not do so, and the problem has vexed scientists and engineers ever since.

DIRECT CURRENT (DC). In a DC circuit, electrical energy flows in one direction only; its polarity is constant. Batteries produce direct current, as do rotary generators with commutating devices. Most of the early practical uses of electricity (telegraphs, telephones, and electroplating) used direct current. These applications informed Edison's personal experience and the published electrical literature on which he drew to design his lighting system, which used DC. After **alternating current** (AC) emerged in the mid-1880s as a feasible option for light and power, DC was sometimes contrasted as the "continuous" or "straight" current. DC's advantages over AC included its technical and conceptual simplicity (it could readily be described by analogies to water flowing in pipes) and the easy conversion of its generators into motors. Its low-voltage distribution also gave it a safety advantage but handicapped it economically against AC. Edison engaged in the **War of the Currents** to defend his DC system against competition from AC.

DREXEL, MORGAN & COMPANY. This **New York City** banking house played key roles in financing Edison's electric lighting ventures and served as Edison's personal bank. Created in 1871 by Anthony Drexel, Junius S. Morgan, and **J. Pierpont Morgan** (son of Junius), it was from the start one of New York's most important banks.

Drexel, Morgan had deep ties to Edison's lighting enterprises. Among its partners who became directors of Edison companies were Egisto Fabbri, James Hood Wright, and **Charles Coster**. The Italian-born Fabbri (1828–1894) was instrumental in establishing Edison's electric light and power business in the United States, Europe, and South America. He was the founding treasurer of the **Edison Electric Light Company** and an organizer of the Edison Electric Light Company Ltd of London. Fabbri also helped manage Drexel, Morgan's interest in the Edison Electric Light Company of Europe. His financial failure in 1883 led to him being replaced by Coster. Wright (1836–1894), a founding partner of Drexel, Morgan, was a trustee of the Edison Electric Light Company and a director of the **Edison Electric Illuminating Company of New York**. Drexel, Morgan's affiliated houses in London and Paris also facilitated the transfer of patent rights and payments for Edison lighting businesses in Europe and Great Britain.

Edison came to believe that Drexel, Morgan was too cautious about capitalizing his U.S. companies' expansion to meet growing markets. Regretting the control he had allowed the firm, he created and financed the **Edison Construction Department** to build central station lighting plants in the United States, For similar reasons, he also underwrote his own factories: the **Edison Machine Works** and the **Edison Lamp Company**.

Edison tried to limit the firm's participation in the consolidation that created the **Edison General Electric Company** (EGE) in 1889. However, Drexel, Morgan ended up with enough leverage to force EGE into a merger with the **Thomson-Houston Electric Company** in 1892. J. P. Morgan and Coster led that effort and largely blocked Edison's allies from taking roles in the new **General Electric Company**.

Despite his frustrations with Drexel, Morgan policies, Edison relied on the firm for his personal banking. It handled his checking and investment accounts, traded in stocks and bonds, processed international payments, and transferred funds to Edison associates and family members abroad.

Drexel, Morgan also exerted great influence on U.S. railroads, especially after the Panic of 1873. By specializing in reorganizing rail lines, it reached a preeminent position among the nation's investment banks. In 1895, after the deaths of the senior Morgan, Drexel, and Wright, the firm reorganized as J. P. Morgan & Company.

DUPLEX TELEGRAPH. Duplex telegraphs enabled two messages to be sent simultaneously over a single wire in opposite directions. A less common method known as diplex sent two messages in the same direction. Duplex telegraphy became a subject of significant interest in the United States following the Civil War as a way to increase carrying capacity without a proportional investment in expensive new lines. In 1865, inventor Moses Farmer demonstrated a duplex system between **Boston** and Portland, Maine, and on a line from Cincinnati to Indianapolis. Three years later, Joseph Stearns successfully introduced a duplex on the lines of the Franklin Telegraph Company in Boston. The **Western Union Telegraph Company** adopted an improved version of the Stearns duplex in 1872. That action prompted Edison, who had experimented with duplex and diplex telegraphy between 1865 and 1869, to resume his research on the subject. In early 1873, Western Union president **William Orton** agreed to allow Edison to test his instruments over the company's wires and to have his instruments made at the company's **New York City** shop supervised by **George Phelps**. Edison in return gave Western Union the right to purchase any of his patented duplex designs. Edison's experiments led him to develop his **quadruplex telegraph**.

DYER, FRANK LEWIS (1870–1941). A longtime Edison business associate, Dyer held several administrative and executive positions. He was an attorney, partnering with his brother **Richard Dyer** from 1897 to 1903. Dyer moved his office near Edison's **laboratory** in **West Orange, New Jersey**, in 1903 to organize Edison's Legal Department, and for the next nine years he was general counsel for various Edison companies. In 1908, he replaced **William Gilmore** as president of the **National Phonograph Company**, vice president of the **Edison Manufacturing Company**, and general manager of the **Edison Phonograph Works**. He also took the lead in organizing the **Motion Picture Patents Company**, serving as its first president. Dyer was a guiding force behind the formation of **Thomas A. Edison Incorporated** in 1911 and its president until 1912, when he resigned. He cowrote (with **Thomas C. Martin**) an official biography and was a member of the **Edison Pioneers**.

DYER, RICHARD NOTT (1858–1914). Dyer was Edison's principal patent attorney for about twenty years, starting in 1882. A graduate of Georgetown University, he took on that role when his father, George Dyer, and **Zenas Wilber** dissolved the partnership that had handled Edison's patent work for several years. For the next twenty years, Dyer, first by himself and then from 1884 with a succession of partners (including his brother **Frank Dyer**), was responsible for preparing, filing, and prosecuting the applications that defined Edison's intellectual property under the U.S. **patent system**. He collaborated with attorneys abroad regarding foreign patents and served as patent counsel to the **Edison General Electric Company** and the **National Phonograph Company**.

E

EATON, SHERBURNE BLAKE (1840–1914). A prominent New York attorney, Eaton filled important roles in Edison's lighting business and for Edison personally. He became vice president and general manager of the **Edison Electric Light Company** in 1881 and its president and de facto manager in 1882. Eaton remained president until October 1884, when he was forced out due to Edison's dissatisfaction with the company's policies. Eaton remained with Edison Electric as its general counsel and took a lead role in patent litigation cases through the 1880s. Until late 1884, he was also vice president of both the **Edison Electric Illuminating Company of New York** and the Edison Electric Light Company of Europe (a patent-holding company in **New York City**). When Edison General Electric was organized in 1889, Eaton became its general counsel. Edison also retained his services as personal attorney from January 1889 through 1892, receiving his advice on a range of affairs.

ECKERT, THOMAS THOMPSON (1825–1910). Eckert was an important telegraph official who acted as something of a foil to the young Edison's ambitions. Eckert had extensive experience as an operator and manager by 1861, and when the Civil War broke out, he was appointed supervisor of the military telegraph office in Washington, D.C. For his service to the president and secretary of war, he was brevetted major general and named assistant secretary of war. He left government service in 1867 and became, in turn, superintendent of **Western Union**'s eastern division, president of the **Atlantic and Pacific Telegraph Company** (A&P) (owned by **Jay Gould**), and president of the American Union Telegraph Company (also organized by Gould).

At the end of 1874, not long after Gould acquired A&P, he, Eckert, and Eckert's assistant Albert Chandler came to Edison's shop in **Newark, New Jersey**, to see the **quadruplex** telegraph, which allowed the simultaneous transmission of four messages on a single wire. Gould purchased the rights to the invention for $30,000, preempting Western Union and precipitating a legal battle that ended only when Western Union absorbed A&P three years later. In 1875, when Edison was A&P's electrician, his relations with Eckert, company president, deteriorated. Edison thought that Eckert was stingy with research and development funds and deliberately impeded technological improvements, and he blamed Eckert for his difficulties with the company. After Eckert returned to Western Union in 1881 as vice president and general manager, Edison reportedly (according to the recollections of **Alfred Tate**) refused to deal directly with him.

EDISON (THOMAS A.) CONSTRUCTION DEPARTMENT. Edison used this organization to build incandescent electric lighting systems in American towns and small cities from May 1883 to the autumn of 1884. An informal financial and administrative framework, the Construction Department had no legal standing and was not part of any company, though Edison coordinated its work with the **Edison Electric Light Company** and the **Edison**

Company for Isolated Lighting. Samuel Insull managed its finances and daily affairs.

Edison wanted the Construction Department to promote his **village plant** system more aggressively than had the Edison Electric Light Company. The strategy was to encourage local investors to form illuminating (utility) companies in promising towns and small cities. Edison surveyed energy use in each place and, if satisfactory, used the Construction Department to build (or subcontract) for the generating plant and distribution network, wire local business and homes, train employees, and operate the station for a short period before turning it over to the utility. This arrangement allowed Edison to add a profit margin to the cost of each plant and created work for his own factories, although he had to advance most expenses from his own pocket. The earliest contract was for Shamokin, in east-central Pennsylvania, but the first finished plant opened in nearby **Sunbury, Pennsylvania**, on 4 July 1883. While the Construction Department's work spread over twenty states, all but two of its sixteen finished plants were in Ohio, Pennsylvania, or Massachusetts.

Even the relatively economical village plant system cost more money to build than most local utilities could raise, dimming the Construction Department's prospects. Edison had already decided to quit the business when a banking panic in May 1884 virtually froze the American financial system. After transferring some staff members to enable work to continue, he handed over operations to the Edison Company for Isolated Lighting on 1 September 1884.

EDISON AND FORD WINTER ESTATES. This twenty-one-acre property in **Fort Myers, Florida**, includes Edison's winter home ("Seminole Lodge"), a guesthouse, his research laboratory, and extensive gardens. It also contains a second home built concurrently with Edison's by his friend **Ezra Gilliland**; it passed to other owners before Henry Ford acquired it in 1916. In 1928, Ford moved the original laboratory building to Dearborn, Michigan, and built a new one for Edison's research on **rubber**. **Mina Edison** deeded the Edison property (including furnishings) to the City of Fort Myers on her death in 1947, and it opened to the public that year. The Ford property ("The Mangoes") was sold in 1945 and obtained by the City of Fort Myers in the late 1980s. The combined Edison–Ford estate opened to the public in 1990. *See also* MENLO PARK LABORATORY IN GREENFIELD VILLAGE.

EDISON COMPANY FOR ISOLATED LIGHTING. This company exemplified the complex relationship between central stations and smaller freestanding generating plants as electric incandescent lighting took hold in the 1880s. As the name suggests, the Isolated Company (as it was called) sold and installed freestanding plants. It began as a bureau within the **Edison Electric Light Company**, but the volume of its business led to it being spun off in late 1881 as a stock company controlled by Edison Electric Light. (Edison himself had no direct financial interest.) The Isolated Company operated in areas not served by either a central generating plant or a municipal gas company, though it competed with isolated **gas lighting** plants that made illuminating gas on-site for single houses or buildings. With permission from Edison Electric Light, however, it also did extensive business in cities with gas service (such as **New York City**) that were obvious markets for future central stations. Its customers included shops, hotels, theaters, factories, mills, and homes. Among its projects was the Southern Exposition of 1883 in Louisville, Kentucky, the largest installation of incandescent lights to date. Easier and cheaper to set up than big central stations, isolated plants spread the reach of Edison's light and created demand for lamps, dynamos, and other products of Edison's factories. Isolated Company agents helped organize local Edison illuminating (utility) companies. The company took over the central station business entirely in September 1884 when the **Edison Construction Department** folded. In another reshuffling in 1886, the Edison Electric Light Company absorbed the Isolated Company and transferred its functions to a new entity: the **Edison United Manufacturing Company**.

EDISON EFFECT. This name was given to a puzzling phenomenon related to the blackening of Edison's early incandescent lamps. The darkening was due to a film of carbon on the glass, which Edison hypothesized resulted from "carrying by Electrification of the Carbon from one side of the Carbon horseshoe" (*TAEB* 5:627). Trying to prevent the "carrying" in February 1880, he created an experimental lamp with a third wire (electrically separate from the filament) inserted through the globe into the space near the filament. The third wire did not prevent "carrying," but Edison found that it had an electrical charge when the lamp ran at high incandescence. When he connected it to a galvanometer (for measuring current), he found a current that rose in a steep but predictable curve as the lamp became brighter. Although unable to explain its cause, Edison saw this current as a useful indicator of voltage on the lamp filament. In October 1883, he used the special lamp in a new form of voltage indicator, a critical need in his **electric lighting system**.

Edwin Houston brought the phenomenon to the attention of physicists and engineers in October 1884. In early 1885, British electrical engineer **William Preece** referred to the "Edison Effect" in a scientific paper, and the name stuck. Almost two decades later, John Ambrose Fleming used a modified Edison Effect bulb to detect radio waves. His "thermionic valve"—based on thermionic emission, or the flow of charged particles from a hot conductor to a cooler one—was the first practical vacuum tube, and it opened the electronic age.

EDISON ELECTRIC ILLUMINATING COMPANY OF NEW YORK. This utility company was formed in December 1880 to provide electric incandescent light and power in Manhattan (the whole of **New York City** at the time). Edison built and operated the **Pearl Street** central station—the first full-scale commercial use of his lighting system—under the company's auspices. The Illuminating Company licensed patents from the **Edison Electric Light Company**. The two companies also shared offices at 65 Fifth Avenue, and their investors and executives overlapped closely.

Edison and other longtime directors left the board in 1899 as the Illuminating Company began to merge with another utility to create the New York Edison Company (1901), precursor of the present-day Consolidated Edison Company (formed 1936).

EDISON ELECTRIC LIGHT COMPANY. This patent-holding firm, the brainchild of attorney **Grosvenor Lowrey**, was incorporated in **New York City** in October 1878 to provide Edison with funds for research and development of his electric light. In return, it owned the resulting U.S. patents, which it licensed to other companies. Its licensees included local Edison utility companies (often in return for stock shares), the **Edison Company for Isolated Lighting** (and its successors), and Edison's own manufacturing shops (**Edison Lamp Company, Edison Machine Works, Electric Tube Company, and Bergmann & Company**). Most of its initial investors had ties to **Western Union Telegraph Company** or **Drexel, Morgan & Company**. Edison Electric Light combined with the surviving Edison shops in early 1889 to create the new **Edison General Electric Company**.

EDISON GENERAL ELECTRIC COMPANY (EGE). In the face of rising competition in the electric light and power industry, EGE was incorporated in April 1889 to consolidate the patent ownership rights of the **Edison Electric Light Company** with the manufacturing capacity of the Edison shops—the **Edison Lamp Company, Edison Machine Works**, and **Bergmann & Company**. (The Sprague Electric Railway & Motor Company and smaller firms were added soon afterward.) EGE reportedly paid $3.5 million in cash and stock for the shops. Edison personally received about $530,000 in cash and almost twice that in stock but would later lament the loss of his steady income from the shops. **Henry Villard**, who worked closely with Edison to create EGE, became its first president, but its day-to-day business was managed by second vice president **Samuel Insull**. Villard brought in fresh capital, notably from **Drexel, Morgan & Company**, Deutsche Bank, and other German investors. Villard set an ambitious course, hoping to create unified systems of electric light, power, and railroads in major cities. He also courted the **Thomson-Houston Electric Company**, a smaller but more profitable rival. In early 1892, facing EGE's unsated need for cash and Villard's personal liquidity problems, both Edison and Villard reluctantly accepted a merger with Thomson-Houston engineered by Drexel, Morgan & Company. The resulting entity did away with personal names altogether. The incorporation of the **General Electric Company** in April 1892 marked the departure of Edison from the electric lighting business and of Samuel Insull from the Edison milieu.

EDISON LAMP COMPANY. One of Edison's three main manufacturing shops in the electric light and power business, the lamp company made incandescent lamps under license from the **Edison Electric Light Company**. Organized first as the Edison Lamp Works in 1880, the Edison Lamp Company was formed in 1881 as a partnership among Edison, **Charles Batchelor**, **Francis Upton**, and **Edward Johnson**; it was incorporated in 1884. It initially operated a small factory adjacent to Edison's **laboratory** in **Menlo Park, New Jersey**, before moving in 1882 to Harrison, New Jersey (adjacent to **Newark, New Jersey**), where it had larger facilities and a pool of cheap labor. Worried by quality control problems, Edison moved his laboratory to the factory in 1886 and kept it there until late 1887. Upton helped to finance the factory's expansion and, as general manager, oversaw its operations for about fifteen years. By the mid-1880s, the profits on its immense production were so great that Edison later recalled, only somewhat in jest, that the Lamp Company paid a dividend every week. The company (with its roughly 500 workers) was absorbed, along with the other shops, into the new **Edison General Electric Company** (EGE) in 1889. It briefly retained an independent identity before becoming the Lamp Manufacturing Department of EGE and then the Lamp Works of the **General Electric Company**.

EDISON MACHINE WORKS. The Machine Works manufactured dynamos, motors, and other heavy equipment for the Edison light and power system. It was also a center for product design and testing (including electrical insulation) and for training electrical workers. Edison formed the Machine Works in 1881 in partnership with **Charles Batchelor**, who took a 10 percent interest. They leased a former ironworks on Goerck Street (now Baruch Place) in Lower Manhattan near the East River. The Machine Works thrived on orders from the **Edison Company for Isolated Lighting**, the **Edison Construction Department**, and central station projects. It survived a financial scandal that led Edison to fire its superintendent and bookkeeper in 1883; it was incorporated the next year, with Batchelor becoming general manager. When a strike over hours, pay, and rules stopped production in May 1885, Edison and Batchelor dropped their plan to expand in **New York City** and instead sought a larger factory and more pliant workers out of the city. They soon bought an abandoned locomotive plant in Schenectady, New York. **John Kruesi** took over as superintendent in 1886 (soon after the Machine Works absorbed the **Electric Tube Company**), and **Samuel Insull** became secretary and treasurer. (Insull later replaced Batchelor as general manager.) The same year, Edison arranged for the Machine Works to build and test railway motors for his former protégé **Frank Sprague**. The Machine Works employed about 900 men when the **Edison General Electric Company** absorbed it in 1889.

EDISON MANUFACTURING COMPANY. Edison created this personal business entity in December 1889 to manufacture and market the Edison–Lalande **battery (primary)**. He had earlier signed a licensing agreement with Felix Lalande to market the battery, which he then improved. Edison Manufacturing built a factory at Silver Lake, New Jersey, that eventually grew into a large complex and opened a sales office in **New York City**. In the early 1890s, the company sold batteries for the **phonograph**, the **phonoplex telegraph**, telephone systems, and medical uses and produced wax for phonograph cylinders. Later in the decade, it manufactured devices such as fan motors, X-ray equipment, and medical instruments. Edison Manufacturing also served as Edison's motion picture company, producing kineto- scopes and films. In 1905, its motion picture operations moved from Manhattan to a stu- dio in the Bronx. The company was formally incorporated only in 1900. **Thomas A. Edison Incorporated** assumed its assets and prop- erty rights in 1911, and Edison Manufacturing was dissolved in 1926. *See also* SILVER LAKE CHEMICAL WORKS.

EDISON MUSEUM AT EDISON PLAZA.

The Edison Museum in Beaumont, Texas, is a science and history museum featuring inter- active exhibits about the life and inventions of Thomas Edison. A decommissioned sub- station of Gulf States Utilities Company (now Entergy Texas) houses the museum. Former Gulf States chief executive W. Donham Craw- ford gathered many of its 1,400 artifacts and much of its reference library. Crawford's widow bequeathed his collection to the museum in the early 1980s.

EDISON ORE MILLING COMPANY.

The Edi- son Ore Milling Company was incorporated in New York State in December 1879 to use Edi- son's ore separation and concentration inven- tions for extracting gold, iron, and other metals from low-grade ores and tailings. Its capitaliza- tion was $350,000, and incorporators included Edison, **James Banker**, Robert Cutting Sr., **Robert Cutting Jr.**, and **Grosvenor Lowrey**. Edison assigned the company his prior con- tracts for processing gold and platinum tail- ings at Cherokee Flats and two other claims in California. In 1881, the company attempted to extract iron from black sand on beaches in Rhode Island and Long Island. None of these early ventures proved profitable, and in 1887, Edison reorganized the company as a patent-holding firm. It thereafter did no mining- related work in its own name but leased its patents to the Edison Iron Concentrating Company and the **New Jersey and Penn- sylvania Concentrating Works** (both Edison enterprises) to process iron tailings and ore at mines in Humboldt, Michigan; Bechtelsville, Pennsylvania; and Ogdensburg, New Jersey. The Edison Ore Milling Company was offi-

cially dissolved on 2 April 1924. *See also* ORE (GOLD) REFINING; ORE (IRON) MILLING.

EDISON PHONOGRAPH COMPANY.

As Edi- son returned to work on the **phonograph** after a long hiatus, he organized this company in the fall of 1887 to manufacture, promote, and sell the instrument in the United States and Canada. (He subsequently granted the man- ufacturing rights to the **Edison Phonograph Works**.) Incorporated in New Jersey, the firm had offices in **New York City**; Edison held the overwhelming majority of its capital stock of $1,200,000. Its formation appeared to usurp long-standing rights of the **Edison Speaking Phonograph Company**, touching off a fight with the latter's investors. Edison installed his friend **Ezra Gilliland** as Edison Phono- graph Company's only sales agent, a position whose ambiguous allegiances fractured their fraternal relationship in less than a year. The company conducted scant—if any—business. Edison had no phonograph to market until he readied the "perfected" model in June 1888, and he almost immediately sold the company to **Jesse Lippincott** as a way of transferring his rights to the **North American Phonograph Company**. Still, Edison Phonograph survived as a legal entity through North American's bankruptcy and dissolution, and it took title to some Edison patents as late as 1897.

EDISON PHONOGRAPH TOY MANUFAC- TURING COMPANY.

This firm was incorpo- rated in Portland, Maine, in October 1887 to manufacture and sell talking dolls and other phonograph-based toys under license from Edison. Its officers included inventor William Jacques (president) and his associate Low- ell Briggs (treasurer). Edison held a minority of stock, and his secretary **Alfred Tate** was a director. The company had offices in **Bos- ton** and **New York City**. Jacques, an **Ameri- can Bell Telephone Company** electrical expert with a PhD in physics, had created a proto- type by adapting Edison's phonograph to a doll. **Charles Batchelor** made further improve- ments at Edison's **laboratory** in **West Orange, New Jersey**.

The company aimed to profit from the domestic sale of dolls (made at the **Edison Phonograph Works**) and from licensing or selling foreign manufacturing and marketing rights. However, design flaws (especially the dolls' inability to withstand regular use) and manufacturing delays frustrated the company's ambitions. Legal challenges brought by Jacques after Edison forced him out in 1889 compounded its troubles. The firm also faced severe challenges in licensing other companies to make and sell dolls in Europe. Edison was reluctant to allow manufacturing by anyone not under his direct employ, and the status of his European patents was uncertain. Aggravated by the absence of promised royalties, Edison withdrew his license to the company in 1891, and litigation followed. A settlement reached in 1895 required the destruction of toy phonographs stored at the Edison Phonograph Works and the sale of the remaining doll bodies. Liquidation of this property ended the company's practical existence.

EDISON PHONOGRAPH WORKS. This firm was incorporated in New Jersey in May 1888 to produce Edison's phonograph and accessories for domestic and foreign markets. Edison assigned it the manufacturing rights conferred by his patents, in return for which he received 52 percent of its stock ($250,000 in total). Edison served as president; original directors included **Charles Batchelor, Robert Cutting Jr., John Tomlinson,** and **Alfred Tate.** The company began production at a small plant in Bloomfield, New Jersey, already set up by **Ezra Gilliland,** but within a few months, it was operating out of its own larger factory next to Edison's **laboratory** in **West Orange, New Jersey.** It made phonographs, wax cylinders (and for a time the wax itself), numbering machines for the Bates Manufacturing Company, and electrical devices for the Edison Manufacturing Company. In its early days, it also produced the Edison–Lalande **battery (primary).** Fire destroyed the factory in 1914, but Edison quickly rebuilt. **Thomas A. Edison Incorporated** absorbed the Phonograph Works in 1924.

EDISON PIONEERS. Thirty-seven former Edison employees and associates created this social organization in 1918 to honor the inventor and their memories of working with him and each other. They met annually on or near Edison's birthday, often with him in attendance. Membership, open at first only to those who had worked with Edison through 1885, rose in a few years to about 110 men. Eligibility eventually expanded to those whom Edison employed before 1900; later employees, as well as children of deceased members, were admitted on separate terms. A few individuals, including Edison and his family, received honorary memberships. The group's projects included erecting commemorative tablets and a memorial tower at **Menlo Park, New Jersey;** gathering artifacts (many of which went to **Henry Ford** for his museum in Dearborn, Michigan); and compiling biographies and obituaries. The Pioneers merged into the Edison Foundation in 1957.

EDISON PORTLAND CEMENT COMPANY (EPCC). Edison created this firm to acquire his processes and patents for manufacturing cement in the United States and Canada, and to produce and sell Portland cement. It was incorporated in New Jersey in June 1899. His venture into cement was an outgrowth of the unprofitable Ogdensburg ore milling operations of the **New Jersey and Pennsylvania Concentrating Works** (NJPCW). In 1898, Edison and NJPCW vice president **Walter Mallory** shifted their attention to adapting Ogdensburg machinery and processes to making cement.

Limestone-based Portland cement, patented in England in 1825, was (and remains) the most common form of cement. Edison, after visiting plants in Pennsylvania and investigating the industry generally, decided to produce it by building a mill in Stewartsville, New Jersey. The area had an abundance of limestone and cement rocks, a mixture of which (known as "clinker") is an essential ingredient.

Edison and a group of Philadelphia investors formed the EPCC with a capital stock of $11 million to finance construction. (Edison received $9 million in common shares and $1 million in preferred stock.) Harlan Page, one of the investors, became president; Mallory was vice president, and Edison was general manager. The company opened sales offices in **New York City**, **Boston**, Philadelphia, and several southern states.

Construction at Stewartsville began in 1900 and took about two years. The plant incorporated major features from Ogdensburg, including the automation of most operations, fine grinding of stone, and improved quarrying methods. Edison also created new equipment and processes, including machinery for automatic sampling, averaging, and mixing the proper proportion of limestone and cement rock. He cut the time for aging cement from months to hours by placing it in a temperature- and humidity-controlled storehouse. And he devised a long rotary kiln to roast the finely crushed limestone and rock. Standard kilns produced up to 200 barrels of clinker a day, but by 1906, each of Edison's six colossal kilns (150 feet long) could turn out five times as much. Other cement companies subsequently adopted the long rotary Edison kiln, and it became the industry standard. The Stewartsville plant expanded several times, employing some 800 workers by 1906. By then, the plant and the workers' housing around it was a separate entity known as New Village.

Edison received forty-nine patents in cement production, some on innovations he developed—but did not patent—for milling iron ore. In addition to the rotary kiln, his patents covered sifting screens; crushing rolls; automated apparatus for sampling, averaging, and mixing; and the method and apparatus for aging the clinker. The EPCC combined the patents in 1908 into a manufacturers' pool led by the North American Portland Cement Company.

The Edison firm had trouble making a profit despite its efficient production and high-quality product. High costs for experimentation and construction (and rebuilding the plant after a March 1903 fire), along with increasing competition, were partly to blame. So too was Edison's insistence on grinding the ingredients to an exceptionally fine state to produce a superior cement. Although the company was

not profitable until 1922, it did become one of the largest producers in the United States. Its cement went into major projects like the original Yankee Stadium, which opened in the Bronx in 1923. The company was dissolved soon after Edison's death in 1931. The Edison Cement Corporation, its successor, operated until 1942. *See also* CEMENT HOUSE.

EDISON SPEAKING PHONOGRAPH COMPANY (ESPC). The ESPC was incorporated in Connecticut in April 1878 to manufacture and sell Edison's tinfoil **phonograph** in the United States (except for use in clocks, watches, or toys). It was also a vehicle for carrying out Edison's February 1878 agreement with a small group of investors who sought to control patent rights to the phonograph. Those suitors included entrepreneur **Gardiner Greene Hubbard** (father-in-law of **Alexander Graham Bell**), and the negotiations took place amid a larger struggle for control of the new telephone industry. ESPC, in return for exclusive manufacturing and sales rights, was to give Edison money to improve the phonograph and a 20 percent royalty on all sales. Edison shared his royalties with **Uriah Painter** (who had brought Hubbard into the deal and became ESPC's largest shareholder) and **Charles Batchelor**. **Edward Johnson** became general sales agent. ESPC offered several phonograph models (some made by a predecessor of **Bergmann & Company**), but with a minimum price of $80 each, Johnson struggled to generate sales. The company focused instead on public exhibitions and hired James Redpath in May 1878 to manage its exhibition business.

By November 1878, Edison, Johnson, and Painter were so dissatisfied with the company's management that they took control and gave day-to-day operations to Johnson. They also brought in **Josiah Reiff** as an investor. Public enthusiasm for the phonograph waned, and ESPC was moribund by the end of 1879. The company lay out of sight until Edison created the **Edison Phonograph Company** in 1887 to bring his new wax-cylinder phonograph to market. That move effectively blocked ESPC from any future phonograph business, and Painter fought back. The dispute frayed Edison's friendships with Johnson and **Sigmund Bergmann**. It also threatened to sink the **North American Phonograph Company**—which had bought the contested rights from Edison Phonograph—until **Jesse Lippincott** agreed in essence to buy out Painter and ESPC.

EDISON STAR. The notion that Edison was responsible for the appearance of a bright object or "star" in the night sky was an enduring popular myth of uncertain origin. Possibly it began around Edison's electric lighting of **Menlo Park, New Jersey**, in 1879–1880, though it may have started about 1885, when he was flying (unilluminated) balloons for experiments on the **railway telegraph**. Edison first fielded questions about it in 1887. The myth outlived his work on electric lighting, and both he and newspaper editors routinely received public inquiries for decades. Edison ascribed the "star" to the bright planet Venus, and in 1927, the astronomer and science writer Garrett Serviss noted a correlation between reports of the "star" and the planet's prominence. However, no amount of denial or explanation could fully extinguish the myth, and reports of the "star" outlived Edison himself. Shortly after his death, residents of **Fort Myers, Florida**, proposed renaming Venus in the inventor's honor.

EDISON STORAGE BATTERY COMPANY (ESBC). Edison organized the ESBC in May 1901 to develop, manufacture, and sell his promising new alkaline **battery (storage)** for electric vehicles. Edison was its president, **Walter Mallory** vice president, and **John Randolph** secretary and treasurer. Capitalized initially at $1 million, the company increased its capital to $3.5 million in 1910 and $5 million in 1917, largely to reduce its indebtedness to Edison, who financed much of its research. Commercial manufacture of a nickel-iron battery began in January 1903 at Glen Ridge, New Jersey, but was suspended in November 1904 when the "E" cells suffered leakage and diminished electrical capacity. Manufacturing resumed only in 1909 (with the "A" cell) after improvements to the battery design and production processes. Edison later expanded

production in a new factory across the street from his **laboratory** in **West Orange, New Jersey**. ESBC's batteries provided power to cars and delivery trucks, mining lamps, train lights and signals, boats, and isolated farms. They also functioned as backup for generating stations. Chemicals for the batteries came from the **Silver Lake Chemical Works**.

ESBC's extensive testing requirements were met initially by the battery research department at the West Orange laboratory and later in the small-cell test department. Yet the mounting workload led to the formation of a separate testing department within the battery factory. In 1913, a research department was created at Silver Lake; enlarged in 1917, it competed with the West Orange laboratory for chemical testing work, including analysis of **batteries (primary)** and Amberola **phonograph** cylinders. ESBC had its own sales force but also sold its products through separate sales companies. In 1932, **Thomas A. Edison Incorporated** absorbed ESBC as its Storage Battery Division. The division manufactured batteries for commercial and industrial uses until the Electric Storage Battery Company (now Exide Technologies) acquired it in 1960.

EDISON UNITED MANUFACTURING COMPANY. A stock company incorporated in 1886, Edison United Manufacturing coordinated the sales and installation of equipment for central stations and isolated lighting plants made by the Edison electrical factories (**Edison Machine Works, Edison Lamp Company,** and **Bergmann & Company**). Its formation allowed the **Edison Company for Isolated Lighting** to exit the central station business. In 1889, Edison United was liquidated shortly after a new company, called the United Edison Manufacturing Company, took over its duties. A year later, the **Edison General Electric Company** absorbed United Edison Manufacturing as its Light and Power Department.

EDISON, CHARLES (1890–1969). The second child of Edison and **Mina Miller Edison**, Charles emerged as the heir apparent to his father's business enterprises. He was a middling student and did not complete his senior year at the Massachusetts Institute of Technology in 1912–1913, but he was mechanically inclined. Because of that talent—and despite an affinity for bohemian **New York City**—he began working for his father in early 1914. Within three years, he became chairman of the board of **Thomas A. Edison Incorporated** (TAE Inc.), which he and **Steven Mambert** reorganized into quasi-independent divisions. Charles took on the roles of general manager (1919) and president (1926). These titles belied his father's power in the organization and his willingness to second-guess or overturn his son's business actions, particularly the progressive employment practices and enlarged workforce after World War I and the decision to enter (belatedly) the consumer radio market. Charles also held executive positions in other Edison-owned firms, including the **Edison Storage Battery Company** and the **Edison Portland Cement Company**. An extended leave from TAE Inc. included stints as assistant secretary of the navy (1937–1940), secretary of the navy (1940), and governor of New Jersey (1941–1944). He did not seek reelection and returned to TAE Inc., remaining there (and at the successor McGraw-Edison Company) until he retired in 1961. He married Carolyn Hawkins in 1918; they had no children.

EDISON, CHARLES PITT (1860–1879). The son of **William Pitt Edison** (Edison's brother) and his wife, young "Charley" grew up in **Port Huron, Michigan**, and began assisting with experiments at his uncle's new **laboratory** in **Menlo Park, New Jersey**, in early 1876. He helped particularly on the **electromotograph telephone receiver** in 1878–1879. Edison sent him and six of the new instruments to London in early 1879 to help **James Adams** (Edison's technical representative in Great Britain) introduce them to the scientific community and prospective investors. Adams died soon after, leaving Charley to make a crucial demonstration at the Royal Society of London. He did so but clashed with **George Gouraud** (Edison's business representative) and abruptly decamped to Paris. Ignoring cables for him to return home, Charley stayed and helped put up **quadruplex telegraph** lines for an American

promoter peripherally connected with Edison. He suffered acute peritonitis and died in Paris in October, cutting short what appeared to be a promising inventive career.

EDISON, MADELEINE. *See* SLOANE, MADELEINE (EDISON).

EDISON, MARION ESTELLE. *See* OESER, MARION ESTELLE (EDISON).

EDISON, MARION WALLACE. *See* PAGE, MARION WALLACE (EDISON).

EDISON, MARY STILWELL (1855–1884). Edison's first wife, Mary was born in **Newark, New Jersey**, to Nicholas Stilwell (a sawyer and occasional inventor) and Margaret Crane Stilwell. The facts of her courtship and marriage were obscured by mythology when Edison became famous, but she seems to have met

Mary Stilwell Edison, 1871. *Courtesy of Thomas Edison National Historical Park.*

Edison by chance in the spring of 1871. She worked briefly at one of his fledgling companies that fall, and they married on Christmas Day 1871. Mary and Edison had three children: **Marion Estelle Edison [Oeser]**, **Thomas Edison Jr.**, and **William Leslie Edison**. Edison privately lamented that his young wife "Dearly Beloved Cannot invent worth a Damn" (Israel 1998, 75), but by all appearances, their marriage was affectionate, and her household management doubtless enabled his long working hours. Several of her relatives worked for Edison, and a sister lived with them; after Mary's death, Edison financially supported members of her extended family. The cause of Mary's early and unexpected death is unknown, but some circumstantial evidence suggests an overdose of medicinal morphine, which was commonly used at the time. She was buried in Newark.

EDISON, MINA MILLER (1865–1947). Edison's second wife, Mina was born into an industrious and devoutly Methodist family in Akron, Ohio, the seventh child of **Lewis Miller** and **Mary Valinda Miller**. She was educated in the Akron public schools, an elite Boston finishing school, and by a European trip with family members. Introduced to Edison by **Ezra Gilliland** and his wife sometime in the first half of 1885, she was engaged that fall and married in early 1886. Mina's much older husband came with the burdens of his famous name, obsessive work habits, and three children. She tolerated all these (and his religious skepticism) and made some early attempts to participate in her husband's inventive work, but most of her energy went to managing the **Glenmont** household in **Llewellyn Park**. In later years, she became deeply involved in the **Chautauqua Institution**, which her father had cofounded. She and Edison had three children, but her relations with the stepchildren were always uneasy (**Marion Edison [Oeser]** was just eight years younger). After Edison's death, she married Edward Everett Hughes. After Hughes died, she resumed using the Edison name. Mina was reburied alongside Edison at Glenmont.

Mina Miller Edison, 1885. *Courtesy of Thomas Edison National Historical Park.*

EDISON, NANCY ELLIOTT (1808–1871). Edison's mother, Nancy, married **Samuel Ogden Edison Jr.** in 1828. They had seven children, of whom Thomas was the youngest. Nancy took primary responsibility for educating Thomas after his brief formal education ended. A former teacher, she was well equipped for the job and could spend much time with him after the last of her other surviving children left home in 1855. She was the daughter of a minister and herself a churchgoing woman who wished to impart a moral sensibility to her son, but their expansive reading together went far beyond Christian teachings. She shaped her son's intellect and, in Edison's words, "taught me how to read good books quickly and correctly" (Israel 1998, 7). *See also* EDUCATION.

EDISON, SAMUEL OGDEN, JR. (1804–1896). Edison's father, Sam, was an independent-minded jack-of-all-trades who prospered as a merchant and opportunistic entrepreneur. Born in Nova Scotia and raised in Ontario, Sam fled Canada to Detroit in 1837 after a brief rebellion against British authorities, for which he faced a treason charge. Following economic opportunities, Sam and his wife **Nancy Elliott Edison** moved their family to **Milan, Ohio**, in 1839 and to **Port Huron, Michigan**, in 1854. Sam's political heterodoxy was reflected in the home library that Nancy used to educate their son, with books by Enlightenment historians and political philosophers such as Edward Gibbon, David Hume, and, above all, the free-thinking religious skeptic **Thomas Paine**. Sam also imparted to his son a belief in abstemious eating as an aid to long life; Edison recalled leaving the table while still hungry as a boy, and he practiced similarly strict dietary habits in his later years. In 1875–1876, when Edison wanted to leave **Newark, New Jersey**, his father helped him find property in nearby **Menlo Park, New Jersey**, and then supervised construction of the new **laboratory** there. (Sam was by that time widowed and romantically involved with his housekeeper, a woman five decades younger, with whom he would have several children.) Sam and a friend embarked on a three-month tour of Europe in 1885, a trip encouraged and paid for by Edison. Sam died in Norwalk, Ohio, and was buried in Port Huron. *See also* EDUCATION.

EDISON, THEODORE MILLER (1898–1992). The last child born to Edison and **Mina Miller Edison**, "Ted" was named for his late uncle Theodore Westwood Miller, recently killed in the Spanish-American War. He was educated at the Haverford School in Pennsylvania and the Montclair Academy in New Jersey, after which he applied his scientific inclination to his father's **World War I research**. Theodore graduated from the Massachusetts Institute of Technology (MIT)—the only Edison child to finish college—after an extra year to complete a master's degree in physics. In 1925, he married Ann Osterhout, whom he met at MIT; they had no children. Obliged to join his father's businesses, Theodore worked at **Thomas A. Edison Incorporated** until 1931, when he started his own engineering firm. He and Ann were active in environmental causes, proponents of zero population growth, and early opponents of the Vietnam War.

EDISON, THOMAS ALVA (1847–1931). "Al," as his family called him, was born on 11 February in **Milan, Ohio**, the seventh (and last) child of **Nancy Elliott Edison** and **Samuel Ogden Edison Jr.** He died on 18 October at **Glenmont**, the **West Orange, New Jersey**, home he shared with his second wife, **Mina Miller Edison**. On the night of his burial three days later, Americans voluntarily turned off their electric lights for a commemorative minute of darkness. Edison was buried in Orange, New Jersey, but in 1963 was reinterred (with Mina) at Glenmont, where the grave site is open to the public.

EDISON, THOMAS ALVA, JR. (1876–1935). Tom Jr. was the second child of Edison and his first wife, **Mary Stilwell Edison**. Edison gave the boy the telegraphic nickname "Dash" to go with "Dot," his older sister **Marion Estelle Edison [Oeser]**. Like his full siblings, Tom was never fully embraced by his stepmother **Mina Miller Edison**, to whom Edison delegated their upbringing. He attended several day schools in and near **New York City** and boarded at St. Paul's School, a prestigious preparatory academy in Concord, New Hampshire, where he was both unhappy and unsuccessful. He did not attend college.

Tom's adult life was marred by ill-health and alcoholism, and the relationship with his famous father was discordant. In 1899, he secretly married stage actress Marie Louise Toohey. That union ended after a few years, and he married Beatrice Heyser in 1906. He had no children. Tom attached his name to a number of dubious business and inventive schemes, antagonizing his father, though he did receive ten U.S. patents. In 1903, he legally agreed to stop using the Edison name in business, and for a time after his remarriage, he called himself Burton Willard, using one of his wife's many pseudonyms. In 1906, with his father's backing, he acquired a farm in Burlington, New Jersey, where he raised mushrooms for more than a decade. Around 1920, he began working at the **laboratory** in **West Orange, New Jersey**, where he developed Edicraft household electric products, such as toasters. The cause and circumstances of his death in a hotel room in Springfield, Massachusetts, remain unknown. *See also* EDISON, WILLIAM LESLIE.

EDISON, WILLIAM LESLIE (1878–1937). William was the third and last child of Edison and his first wife, **Mary Stilwell Edison.** He was born after a difficult pregnancy marked by Edison's extended absence in the West. Like his siblings, he was never fully embraced by his stepmother **Mina Miller Edison,** to whom Edison delegated their upbringing, and his adult relationship with his famous father was discordant. William attended several day schools in and near **New York City** and boarded at St. Paul's School, a prestigious preparatory academy in Concord, New Hampshire. After a brief attendance at Yale University, he enlisted in 1898 to serve as an engineer during the Spanish-American War. He was mechanically inclined, and after the war, he embarked on a series of automotive business ventures, most of them financed by his father. None of them was successful, with the Edison Auto Accessories Company especially aggravating Edison by trading on his name. William turned in 1913 to raising game birds, by which he was able to support himself and his wife (he married Blanche Fowler Travers in 1899; they had no children). He again joined the army in 1918, serving in France. In his last decade, William received five U.S. patents related to radio and signaling systems. *See also* EDISON, THOMAS ALVA, JR.; OESER, MARION ESTELLE (EDISON).

EDISON, WILLIAM PITT (1831–1891). Edison's oldest brother, Pitt (as he was called) was born to **Nancy Elliott Edison** and **Samuel Ogden Edison Jr.** in Vienna, Ontario. He came with the family to **Port Huron, Michigan,** in 1854. When the horse-drawn Port Huron and Gratiot Street Railway began operating in 1866, Pitt became its manager and a substantial stakeholder. He prospered for a time, but over succeeding years, he repeatedly needed help from Edison to make good his investments in the road and various other enterprises, including a farm. Edison hired Pitt's only son, Charles ("Charley"), as an assistant at the **laboratory** in **Menlo Park, New Jersey,** in 1878. Charley went to London in early 1879 to help start the Edison telephone business in Great Britain but contracted an acute infection in Paris and died there in October.

EDUCATION. Edison recalled as an adult that his mother took him out of school as a young boy after a teacher called him "addled." Whether the story has any truth behind it, Edison had only a limited formal education. When he was seven, his family moved to **Port Huron, Michigan,** where he briefly attended the private school of Reverend George Engle. The family's inability to pay the relatively high tuition and other fees likely ended his studies there. **Nancy Edison** then took over her son's education at home. As he later remembered, "My mother taught me how to read good books quickly and correctly, and as this opened up a great world in literature, I have always been very thankful for this early training" (Israel 1998, 7). Nancy likely used standard primers, spellers, and readers to teach her son reading and writing. A religious woman, she would also have employed Bible readings, and Edison attended Sunday school at her church. He also recalled reading books by Enlightenment authors such as Edward Gibbon, David Hume, and **Thomas Paine** from the library of his father, **Samuel Edison.** Edison's homeschooling ended at about age twelve, when he enrolled in the Union School in Port Huron. This school was noted for its curriculum in mathematics, science, and drawing, and it was here that Edison encountered Richard Parker's textbook *A School Compendium of Natural and Experimental Philosophy.*

Much of Edison's subsequent education came through a combination of reading and practical experience. For example, Parker's text contained a detailed discussion of telegraphy, and Edison likely used it in building a half-mile telegraph line from a friend's house to his own. Around the same time, he taught himself chemistry by conducting experiments described in an English translation of a textbook by German chemist Carl Fresenius. Contemporaries remembered Edison as a boy who often made things in his home workshop.

Later, as a telegraph operator, he expanded his knowledge of electricity and related fields by reading standard works (including the *Experimental Researches* of **Michael Faraday**) and by experimenting with telegraph equipment.

Edison's experience led him to develop an educational philosophy highly critical of the typical classical education of the time. He faulted schools for incorporating too many impractical subjects (especially Latin and Greek) when they should have emphasized the practical working out of problems. While books could "show the theory of things," he argued in a 1911 interview, "doing the thing itself is what counts" ("Edison on Invention and Inventors," *Century Magazine* 82:418). He also appreciated the Montessori method, which taught through play and advocated the use of film as a visual teaching tool.

Nonetheless, Edison did not disdain theoretical knowledge and formal education in training engineers. When the lack of electrical engineers threatened to hamper growth of the new electric light and power industry in the early 1880s, Edison turned the testing room of the **Edison Machine Works** into a school (complete with examinations) for men about to run new Edison central stations. He sought a more permanent solution by encouraging the development of electrical engineering programs at several colleges, including Columbia and Cornell, and providing them equipment from the **Edison Electric Light Company**. His laboratories at **Menlo Park, New Jersey**, and **West Orange, New Jersey**, included many men (and only men) who graduated from engineering schools, and many of his chemists had doctorates from German universities. Still, Edison was always willing to hire ambitious young men with practical experience to learn by working in the laboratory. In the 1920s, he created questionnaires for prospective employees that focused on general information, such as that acquired from an encyclopedia, rather than the more specialized knowledge gained through a university education.

Although Edison objected to traditional classical education for boys, he allowed his sons from his first marriage (**Thomas Edison Jr.** and **William Leslie Edison**) to attend St. Paul's School, an elite preparatory institution in Concord, New Hampshire. That decision came from **Mina Miller Edison** (the boys' stepmother), whose younger brothers went there. Like many of his contemporaries, Edison left his children's upbringing—including their education—to his wife. Mina also expected her stepsons, like her brothers, to continue their classical education at Yale. Tom Jr. did not go to college, but William briefly and unsuccessfully attended Yale's Sheffield Scientific School. Circumstances were different when Mina's own sons (**Charles Edison** and **Theodore Edison**, born in the 1890s) were ready for college. Both attended schools that offered scientific coursework in addition to the classical curriculum, and they later enrolled at the Massachusetts Institute of Technology—one of the foremost engineering colleges. *See also* THURSTON, ROBERT HENRY.

ELECTRIC GENERATOR. A generator converts mechanical motion into electric current by electromagnetic induction—the result of moving a conductor through a magnetic field. The word "dynamo" (often used by Edison) refers specifically to a generator with a feedback loop to use its own output for energizing its field magnets.

When Edison began working on his incandescent **electric lighting system** in 1878, he expected to adapt generators used for arc lighting and electroplating, but he found them ill-suited to his needs. They were designed on the principle that their internal electrical resistance should equal that of the outside circuit. That idea came from the fact that a battery produces its maximum current under those conditions, and it held true for generators as well. However, Edison grasped that to make electric lighting economically practical on a large scale, what mattered was not a generator's capacity but its efficiency at converting mechanical energy into electricity. His overarching goal was to produce as much light as possible from each pound of coal, and his generator would play a crucial role in that process. With assistants **Charles Batchelor** and **Francis Upton**, Edison investigated the principles of induction (closely following **Michael Faraday**) and generator design in early 1879. They determined that a small internal resistance (relative to that outside) would maximize the generator's efficiency. That recognition reinforced Edison's determination to design an **electric lamp** on the principle of **high resistance**. Work on these critical components—the generator and the lamp—proceeded in a stepwise, complementary fashion throughout the process of creating an integrated lighting system.

Edison's other key insight concerned the magnetic field. A generator produces current when its armature (carrying conducting wires) moves through the lines of force emanating from a magnet. Drawing on Faraday's work and his own intuition, Edison believed that the more lines of force crossed in the most direct manner, the more productive the generator would be. He thus adopted exceptionally long electromagnets to concentrate Faraday's lines of force into a small space traversed by the armature, giving his machines a unique profile. British electrician **John Hopkinson** later demonstrated that short-core field magnets were superior, and Edison followed suit. The early Edison generators were also easily converted to run as electric motors. The **Edison Machine Works** manufactured generators in a range of sizes for commercial electric systems. *See also* ELECTROMAGNETISM.

ELECTRIC LAMP. Experimenters had been trying for forty years to create incandescent electric lamps when Edison took up the problem in September 1878. The basic challenge was to bring a piece of matter to a temperature high enough to emit useful light without it melting, oxidizing (burning), or consuming an inordinate amount of electric power. Edison had briefly experimented with electric lighting in 1877 but made it a priority only the next year after seeing the arc light system developed by William Wallace and Moses Farmer. Their work seemed to validate the approach he already had in mind (based on the earlier experiments) for an incandescent lighting system. Edison strove to develop a thermal regulator to prevent the incandescing element from overheating and melting. Because the regulator would interrupt the circuit to individ-

ual lamps, he placed lamps in parallel circuits (unlike arc lights, which operated in series) so that each one could be turned on and off without affecting others up or down the line. (Individual operation of lamps would be familiar to users of gas lights.) Edison had already determined in his 1877 experiments that lamps in parallel circuits should have **high resistance** to minimize the amount of current needed. Guided by an intuitive understanding of electrical laws (especially Joule's and Ohm's laws), he made high resistance a priority when he turned full-time to lighting in September 1878. His goal was 100 ohms, which would minimize both the current to be generated and the amount of expensive copper needed in conductors to carry the power.

Starting in September 1878, Edison designed lamps with wire filaments ("burners") of **platinum** because that rare metal has a high melting point. However, months later, systematic basic research into platinum's properties showed that it absorbed air into microscopic pores when heated, weakening the metal and lowering its melting point. Edison consequently tried isolating the filament in a vacuum bulb. The vacuum improved the lamps' performance and durability, but they were still too expensive to make and operate. Platinum was not only costly but also had a low resistance to the electric current.

However, Edison's high-vacuum bulb allowed him to consider carbon as an alternative. Carbon had high resistance but would burn readily in the presence of air. However, the vacuum bulb would preserve it from air. The first newspaper account of his successful carbon lamp described the "eureka" moment when Edison realized he could make carbon into a wire-like filament by using lampblack, the same material in his **carbon-button telephone transmitter**. On 21–22 October 1879, he and his staff made their first successful experiments with a filament of carbonized cotton thread in a vacuum: the carbon-filament incandescent lamp. He filed the fundamental patent application just a few weeks later for the carbon-filament incandescent lamp.

In the days around New Year's Day 1880, Edison showed his lamps to flocks of visitors

Commercial bamboo-filament lamp with socket, 1883. *Courtesy of Thomas Edison National Historical Park.*

to **Menlo Park, New Jersey.** The lamps had filaments of carbonized cardboard bent into a horseshoe shape. Cardboard proved adequate for demonstration purposes but was unsuitable for everyday commercial use. Edison assigned one of his chemists, Dr. Otto Moses, to make a systematic study of scientific literature on carbon substances, and this study guided ongoing filament research. Experiments focused on grasses and canes (such as hemp, palmetto, and bamboo), from whose long, uniform fibers Edison could make sturdy and long-lasting filaments. Bamboo turned out to be best for the commercial lamp. Edison spent years trying to improve the efficiency and durability of his lamp, even as the **Edison Lamp Company** produced millions of them. The iconic pear-shaped bulb became synonymous with Edison and took on broader associations with inventions and new ideas. *See also*

ELECTRIC LIGHT AND POWER SYSTEM; FIBER SEARCH; UPTON, FRANCIS; VACUUM TECHNOLOGY.

ELECTRIC LIGHT AND POWER SYSTEM.

From the start of his intensive work on incandescent lighting in September 1878, Edison aimed to design not only a lamp but also an entire electrical system. He took as his model the **gas lighting** systems used in many cities. These included a central plant from which gas flowed through underground conductors to individual buildings, where it was divided among individual rooms and lamp fixtures that could be turned on an off separately. Each customer also had a meter to measure how much gas should be billed to their account. Edison planned an analogous system in which electricity would flow from central generating plants through underground copper conductors to buildings, where it would be metered and directed to different rooms, where lamps could be turned on or off at will. The copper conductors were the most expensive element, and Edison had to reduce their size to make his system competitive with gas. Unlike most contemporaries, Edison intuitively understood Ohm's and Joule's electrical laws and saw in them a solution to the conductor problem. By adopting lamps with **high resistance**, he could use correspondingly high voltages that minimized the amount of current required—and consequently the size of conductors to carry it (also the amount of energy lost as heat in distribution). Edison recognized that high-resistance **electric lamps** worked best in parallel circuits, which also made it possible to operate each one independently.

Edison created the basic components of his light and power system in 1879 and 1880. Each piece was designed for compatibility with the others according to the criteria governing the entire integrated system. For example, the **generator** came first, tailored to a lamp that did not yet exist but whose electrical characteristics Edison had specified. By subdividing the work of his staff at the **laboratory** in **Menlo Park, New Jersey,** Edison was able to create all elements of a basic **direct current** (DC) distribution system in this brief period. In addition to the iconic lamp, major components included underground conductors, meters, safety fuses, fixtures, and sockets. Edison also experimented with electric **motors** for industries and electric railways. These non-lighting uses of electricity offered the chance to balance the load on central stations and generate revenue over more hours of the day. However, the wide adoption of electric motors would require another decade.

Edison's DC system worked most efficiently in dense urban centers and in isolated plants for individual buildings. His initial central stations (such as **Pearl Street Station** ["First District"] in **New York City**) covered only about a square mile. The **three-wire system** increased the distance over which electricity could be distributed economically, but by the late 1880s, Edison's DC model faced increasing competition from **alternating current** (AC) systems that used higher voltages to carry power efficiently over longer distances. Edison refused to develop his own AC system due to his belief in the inherent danger of high voltages as well as his substantial reputational and financial investments in low-voltage DC technology. The economic and engineering rivalry between DC and AC (particularly between the systems promoted by Edison and **George Westinghouse**) spilled into public view in the late 1880s as the so-called **War of the Currents**. Edison finally began working seriously on alternating current in 1890–1891, but he did not develop a commercially viable system before the Edison company merged into **General Electric** in 1892. At that point, Edison left the industry he had helped to found. *See also* ELECTROCUTION (ANIMAL EXPERIMENTS).

ELECTRIC MOTORS. Like the inverse of an **electric generator**, a motor converts electrical energy into mechanical motion. Among the earliest commercial uses for small electric motors were printing telegraphs for **stock tickers** and private-line printers, including those developed by Edison in the early 1870s. Edison's **electric pen** (introduced in 1875) was one of the first consumer products to have a small electric motor. Although Edison periodically experimented with small motors in the

late 1870s, development of an **electric light and power system** prompted his sustained research on larger and more powerful devices. His first motors were essentially Edison generators (with their distinctively long magnets) with the armature connections reversed. In 1879, Edison adapted motors for sewing machines and for **electric railways**; experiments for the latter continued through 1882. After **Frank Sprague** independently created and began to market an improved motor for industry and railroads, Edison again turned his attention to electric power. Between 1888 and 1891, he and **Arthur Kennelly** designed improved motors for the Sprague system (marketed by **Edison General Electric Company** [EGE]) and for an experimental system powered by low-voltage current carried through the rails. The two men also led a successful effort to create for EGE a series of fractional-horsepower motors for a range of commercial and industrial applications. The versatility of such motors in general, combined with their compactness, cleanliness, reliability, and low operating costs compared with steam engines, led to exponential growth in their use from the early 1890s, and numerous manufacturers entered the business.

ELECTRIC PEN. The electric pen, with an accompanying press, formed a copying system that Edison and the staff of his **laboratory** in **Newark, New Jersey**, developed in the summer of 1875. The pen used a small **electric motor** to drive a slender needle up and down the pen's shaft, enabling a user to create a perforated stencil while writing or drawing. This stencil was then placed in the press, where a roller squeegeed ink through its tiny holes to create a copy of the document. The device was one of the earliest consumer uses of an electric motor. While the need for a **battery (primary)** to drive the motor presented difficulties for users, Edison's electric pen and autographic press sold well for several years in the United States, Canada, Europe, and the British colonies. By the early 1880s, sales fell off, as the pen faced competition from a host of mechanical pens that did not require batteries. One offshoot of the electric pen was the mimeograph developed by **A. B. Dick** in the mid-1880s using another of Edison's patented stencil-making processes. The electric pen itself had a second life in the 1890s after Samuel O'Reilly converted it into the first electric tattoo needle.

ELECTRIC RAILWAY. Edison first conceived of an electric railway in May 1879 but did not experiment with the idea until a year later, when he foresaw the advantages of selling power during the day to help offset the cost of the expensive central station he was planning to build in **New York City**. He built an experimental line in **Menlo Park, New Jersey**, funded by transportation financier **Henry Villard**, owner of the Northern Pacific and an investor in the **Edison Electric Light Company**. This railroad operated until September 1881, when Edison and Villard decided to build a new 2.5-mile line to test electric locomotives for long-distance transportation. Edison's patents were combined with those of inventor Stephen Field in the Electric Railway Company, but that company never built any commercial lines.

Edison returned to electric railways in 1886, when his former assistant **Frank Sprague** demonstrated a new system using overhead wires. The Sprague design proved successful, and its major components were manufactured by the **Edison Machine Works**. After its formation in 1889, the **Edison General Electric Company** (EGE) took control of the Sprague Electric Railway and Motor Company and marketed the system under Edison's name. In the meantime, Edison had begun work on an alternative arrangement that provided low-voltage power to car motors through the rails rather than overhead wires, which some cities had banned for safety reasons. Edison filed his first patent application on his new system in March 1887 and made extensive tests in late 1888 and early 1889 on a horse-drawn rail line near his **laboratory** in **West Orange, New Jersey**, with Villard again providing much of the funding. He resumed experiments in June 1890, by which time numerous American cities were adopting electric railways. Edison and Villard also hoped to use a related system to power electric locomotives for long-distance

freight trains, but the 1892 merger of EGE with **Thomson Houston Electric** into the new **General Electric Company** largely ended Edison's electric railway experiments.

ELECTRIC TUBE COMPANY. To carry electrical current safely underground from central station generating plants, Edison designed a segmented conduit with copper conductors insulated and waterproofed inside a protective tube or pipe. The Electric Tube Company was incorporated in March 1881 to make (and in some cases install) these "tubes." The company's shares were divided equally at the outset among Edison, **John Kruesi, Charles Batchelor**, and two partners of **Drexel, Morgan & Company**. Kruesi managed its factory at 65 Washington Street on the West Side of Lower Manhattan, not far from where he supervised the installation of conductors around the **Pearl Street Station** ("First District") on the East Side in 1881–1882. In 1884, the shop moved to larger quarters in Brooklyn on Bridge Street near the East River. In addition to installing conductors at Pearl Street, the company worked on the handful of underground central station systems built by the **Edison Construction Department**. It also fabricated conductors for stations abroad, including those in London, Paris, and Milan. The Electric Tube Company was absorbed by the **Edison Machine Works** at the end of 1885.

ELECTROCUTION (ANIMAL EXPERIMENTS). Edison's staff at the **laboratory** in **West Orange, New Jersey**, made a grisly series of animal electrocution experiments in 1888 and early 1889. Edison also permitted participation by outside parties whose involvement escalated the **War of the Currents**. Edison is known to have personally participated only twice, but the **Edison Electric Light Company** reimbursed most of the expenses.

These events started when the *New York World* questioned the decision by New York State to adopt **electrocution for capital punishment**. At the paper's request, Edison agreed to provide the practical information needed to carry out a death sentence. (He had recently advised the American Society for the Prevention of Cruelty to Animals—without benefit of experiments—about euthanizing unwanted animals by electricity.) **Arthur Kennelly** and **Charles Batchelor** made the first trials on 21 June 1888. They induced a stray dog, after several attempts, to complete an **alternating current** (AC) circuit. Death followed almost instantly, but a *World* reporter, having watched the dog struggle to free itself, anticipated that electrocution might be prone to mistakes and possibly create "a more hideous spectacle" than hanging (*TAEB* 9:263). However, Edison predicted that electricity would kill a man "in an incalculable space of time. . . . An electric light current will kill a regiment in the ten thousandth part of a second" (*TAEB* 9:264).

Kennelly intermittently made more tests in July, when anti-AC provocateur **Harold Brown** joined him. After reading about Kennelly's earliest experiments, Brown had asked to borrow equipment to make his own trials, but Edison instead invited him to the laboratory. Although Kennelly would later explain that he wanted to find the current needed to kill a dog—implying euthanasia—he clearly had human physiology in mind. After recording the electrical resistance of his and Brown's bodies, he measured the resistance of thirty-nine laboratory employees and 220 workers at the phonograph factory, noting specifically the conditions affecting the passage of current. He also intentionally subjected himself to a severe shock. Kennelly made the animal tests without his usual rigor and struggled to manage numerous variables. These included the animals' size, method of completing the circuit, and combinations of current strength, voltage, and waveforms. Many tests were inconclusive because the shocks did not always cause death—and never instantaneously. Even so, the results convinced him that alternating or pulsating current was more lethal than low-voltage **direct current**, especially as the number of reversals or pulsations increased, as Brown had proclaimed. Kennelly was sufficiently confident to draft an explanatory letter to the *World*, but his experimental inconsistencies did not provide the clarity the paper had sought. However, there can be no doubt that some of the dogs killed (more than a dozen

by Kennelly and others separately by Brown) suffered greatly.

The Medico-Legal Society, charged with advising the state, arranged an additional demonstration at Edison's lab in December. Its committee on capital punishment, aiming to show decisively the power of AC to kill animals as large as a man, arranged for Kennelly and Brown to dispatch two calves and a horse. Edison attended and afterward addressed the witnesses, though his words went unrecorded. The Society voted to recommend AC for execution. At the request of the state's prison superintendent, Edison's laboratory hosted further animal trials to determine the best machinery and methods for that grim purpose. Those final tests on 12 March 1889 used AC to kill four dogs, four calves, and a horse. Fifteen months later, convicted murderer William Kemmler became the first person intentionally electrocuted by the state. *See also* TOPSY.

ELECTROCUTION (CAPITAL PUNISHMENT). Although philosophically opposed to the death penalty, Edison provided technical advice to New York State about using electricity to execute criminals. He reasoned that if the state were to administer death, it should do so as painlessly and humanely as possible, and electricity would kill before the brain could register pain.

Edison's involvement began in late 1887, when Alfred Southwick, a dentist serving on a commission to find a more humane alternative to hanging, contacted him. When the state formally adopted execution by electrocution in June 1888, the *New York World* declared it had done so with no understanding of how to carry out such a sentence. At the invitation of the *World*, Edison agreed to try to supply the necessary practical knowledge. (Coincidentally, the American Society for the Prevention of Cruelty to Animals had recently sought Edison's advice about humanely euthanizing unwanted animals, and he recommended electricity.) In the summer of 1888, Edison authorized laboratory experiments (led by **Arthur Kennelly**) on **electrocution (animal experiments)** aimed at extrapolating the best conditions and procedures to kill a human, and the resulting information was given to a panel advising the state. Edison also granted the wish of **Harold Brown**, a self-taught electrician and crusader against **alternating current** (AC), to participate in the gruesome tests. Brown used the results to bolster his incendiary accusations about the inherent dangers of AC.

Edison also saw the polemical value of the tests, which appeared to show the greater lethality of AC compared with **direct current** (DC). He injected the issue into the **War of the Currents** between his DC system and the AC system of the **Westinghouse Electric Company**, inflaming the rivalry. In advocating for AC as the current best able to cause instant death, Edison went so far as to specify Westinghouse equipment for the New York death chambers. He also testified to the lethal efficacy of AC after William Kemmler, the first inmate sentenced to die by electrocution, appealed the punishment as unconstitutionally cruel and unusual. The court accepted Edison's expert opinion and affirmed Kemmler's fate. As New York State prepared to execute Kemmler, the **Edison Electric Light Company** helped Brown make a surreptitious purchase of Westinghouse dynamos on behalf of the state prisons. Despite the planning and Edison's assurances, Kemmler's execution in August 1890 was a horror: it required more than one application of the AC current over an agonizing period, sickened spectators, and inspired sensational newspaper stories. New York nevertheless remained committed to electrocution, and many other states followed its lead.

ELECTROMAGNETISM. As the name suggests, this term pertains to physical manifestations of the interrelationship between electricity and magnetism. Much of Edison's work, especially in the first half of his life, was based on the conversion of electrical energy into magnetism or back again—in other words, on electromagnetism. He invented, improved, and used countless electromagnetic devices in telegraphy and electric lighting. Examples include **electric generators** and **electric motors**, which are based on the phenomenon

of induction—the production of current in an electrical conductor by a changing magnetic field or vice versa. Electromagnetism was a young science when Edison began his career, and its theoretical basis set out in the 1870s by **James Clerk Maxwell** was not well understood for another decade. Edison learned the science in other ways. One was by using telegraph instruments like sounders and sending keys, which had inductive coils. He read technical books and journals, especially the works of **Michael Faraday**, whose experiments extended and systematized the knowledge of his (mid-century) period and laid out a conceptual framework that Edison grasped intuitively. He hired assistants like Francis Upton with formal training in science or engineering. And he experimented himself. Some of his experiments met practical needs, but others were to satisfy his curiosity about the properties of magnets and the force they projected into space. At various times, he tried to quantify the relationship between a magnet's length and its field strength and to correlate an electromagnet's discharge time with other physical properties.

The convertibility of unseen magnetic and electrical energies was a powerful idea in Edison's mind that helped guide his intermittent searches for **unknown natural forces**. At times, he ventured into qualitative speculations, implicitly touching on a long-running scientific controversy over the primacy of matter or force. He did not articulate a coherent viewpoint, but his speculations can be read as a search for unified physical explanations centered on electromagnetic forces. He suggested, for example, that gravitation is an electromagnetic phenomenon, with the implication that electromagnetism could explain celestial mechanics—the workings of the solar system. In putting these unsubstantiated but plausible ideas on paper, Edison was following a path marked by Faraday, who defended such speculations—however "doubtful, and liable to error and to change"— as "wonderful aids in the hands of the experimentalist and mathematician" (*TAEB* 8:484). *See also* ASTRONOMY; MATTER (EDISON'S CONCEPTIONS OF).

ELECTROMOTOGRAPH. Edison used this term to describe the electrochemical phenomenon of varying friction between specific materials by passing a current through them. While seeking the best combination of metallic recording styli and chemical solutions for the transmitter of his **automatic telegraph** system, he discovered that electrochemical reactions could reduce (or occasionally increase) the friction between his recording pen and the chemically treated paper. He considered the effect to be a "new force." Edison periodically experimented with different uses for this effect, including telegraph relays, electrical measuring devices, and telephone receivers He also found that the effect was reversible: the friction in electromotograph devices could generate a considerable electric current. His only commercial application of the electromotograph principle was in his loudspeaking **electromotograph telephone receiver**, which used a rotating cylinder of a chalklike material in place of the chemically treated paper.

ELECTROMOTOGRAPH TELEPHONE RECEIVER. While Edison focused most of his telephone research in 1876–1877 on what would become the **carbon-button transmitter**, he also sought to develop an alternative to **Alexander Graham Bell**'s electromagnetic receiver, which produced weak sounds. Edison decided in March 1877 that "the best way to receive and hear the words is on the **Electromotograph** principle" (*TAEB* 3:270). The receiver used the variable friction of an electrically sensitive material against an electrode to vibrate a diaphragm and transform signals from the transmitter back into sound. The instrument initially worked better with music than speech and was exhibited as Edison's "musical telephone," while commercial Edison telephones used a receiver like Bell's. Edison continued trying to adapt electromotograph receivers for speech, and public exhibitions showcased the volume they produced. After a lull, Edison resumed work on electromotograph receivers in mid-1878, and by October, he and his nephew **Charles Edison** had designed a rotating button of a chalklike material as the friction surface. By February 1879, the receiver had progressed sufficiently for Edison to draft a patent application (U.S. Patent 221,957) and send Charley to England with six instruments. A successful demonstration before the Royal Society of London reassured prospective inventors and led to the formation of the Edison Telephone Company of London to compete against Bell's company and its electromagnetic receiver. At the same time, **Western Union** acquired the U.S. and Canadian rights to the electromotograph receiver. By mid-1879, Edison had produced an instrument for commercial service known as Edison's loud-speaking telephone, which paired the electromotograph receiver with his carbon-button transmitter. However, the receiver did not hold up to daily use, prompting Edison and **Charles Batchelor** to spend much of the summer and early fall experimenting with better compositions of the chalk button and ways to regulate its moisture. Their efforts notwithstanding, the electromotograph receiver was never used commercially in the United States and fell out of favor in Britain after the 1881 merger of the Edison and Bell companies there.

ETHERIC FORCE. *See* UNKNOWN NATURAL FORCES.

EXPOSITION INTERNATIONALE DE L'ÉCTRICITÉ (PARIS, 1881). *See* PARIS INTERNATIONAL ELECTRICAL EXPOSITION (1881).

EXPOSITION UNIVERSELLE (PARIS, 1878). *See* PARIS UNIVERSAL EXPOSITION (1878).

EXPOSITION UNIVERSELLE (PARIS, 1889). *See* PARIS UNIVERSAL EXPOSITION (1889).

FARADAY, MICHAEL (1791–1867). The work and scientific outlook of this English experimenter deeply influenced Edison throughout his life. Faraday's empiricism, lucid prose, highly visual understanding, and avoidance of advanced mathematics all appealed to Edison. Faraday laid out rigorous and systematic inquiries in three volumes of *Experimental Researches in Electricity*, which Edison discovered as an itinerant telegrapher and later began to reproduce on his own. At his **laboratory** in **Newark, New Jersey**, the aspiring inventor styled his own notebooks as "Experimental Researches," sequentially numbering entries as Faraday had done. Beyond this contrived resemblance, Faraday's understanding of energy (and of electricity and magnetism in particular) struck a chord with Edison's mind and intuitions. The "Master Experimenter," as Edison called him, offered economical descriptions of phenomena without resorting to abstruse theories advanced by physicists who represented the world through mathematics (Israel 1998, 96). Faraday's explications of **electromagnetism** fundamentally shaped Edison's understanding. He rendered electromagnetic induction so vividly in plain English and sketches that Edison could clearly envision the fields and lines of magnetic force. Translating them into drawings on paper, Edison designed magnetic circuits for **generators** more advanced than any that could have been derived mathematically at the time. Edison also took to heart Faraday's exquisite sensitivity to different forms of energy and the convertibility of one form to another in seemingly inert space. (Before the word "dynamo" was widely used,

Edison proposed calling a device for changing magnetic and mechanical energy into electricity a "Faradic machine"). Edison's understanding of the phenomena behind the **railway** ("**grasshopper**") **telegraph** and his episodic studies of magnets were grounded in Faraday's work. His long-standing hunches about the existence of **unknown natural forces** emanated from Faraday as well.

Edison arguably learned from Faraday as much about the process of gaining new knowledge as he did about the physical world itself. Faraday's empirical approach—informed questioning, purposeful experiment, meticulous observation, and careful reasoning—created the template Edison used to delve into problems. Even Edison's seemingly fanciful inquiries into new forces and the nature of physical matter followed Faraday's belief that principled speculations, far from being useless, "are wonderful aids in the hands of the experimentalist and mathematician." *See also* EDUCATION; MATTER (EDISON'S CONCEPTIONS OF); MAXWELL, JAMES CLERK; SCIENCE (EDISON AND); THOMSON, WILLIAM (LORD KELVIN).

FESSENDEN, REGINALD AUBREY (1866–1932). Fessenden did valuable chemical work in his relatively brief tenure at Edison's **laboratory** in **West Orange, New Jersey**. A native of Québec, he came to New York via Bermuda and applied to Edison for employment on the strength of his "thorough grounding in mathematics" and desire to learn electrical engineering. However, when Fessenden reported to work in January 1888, Edison assigned him to the chemical laboratory and told him to learn chemistry. Fessenden took charge of the chemical lab after only a few months, when Edison dismissed the academically inclined incumbent. He worked mainly on fireproof insulation for electrical wires, carbon lamp filaments, and copper and nickel ores. Edison cut his laboratory staff by about one-third in April 1890, closing the chemical laboratory and discharging Fessenden. Edison estimated him at the time as a "persistent worker and good Experimenter," but Fessenden was among the Edison alumni to make major advances elsewhere in science or technology (TAE to Fessenden, 26 April 1890). His contributions to the development of radio broadcasting included the first transmission of music and voice (1906) and the principle of amplitude modulation (the basis of AM radio).

FIBER SEARCH. Edison spent years improving the lamp filament of carbonized plant fiber, starting with the raw material. Focused initially on grasses and canes, he is supposed to have predicted that "somewhere in God Almighty's workshop there is a vegetable growth with geometrically parallel fibers suitable to our use" (*TAEB* 5:767). The fibers had to meet other criteria, such as strength, electrical resistance, and availability. The initial search in 1880 resembled his earlier quest for **platinum** in that he directed a review of published literature, sent inquiries to far-flung officials, and dispatched his own agents to investigate locally (in Georgia, Florida, Cuba, Brazil, and Japan). The process resulted in Edison selecting a form of Japanese bamboo for the first commercial lamps, although he continued to experiment with natural and artificial alternatives. He made another big push in 1886–1888 in hopes of increasing the filaments' durability and efficiency while also lowering their cost. At least five hardy travelers embarked for disparate regions, including Mexico, India, Singapore, Colombia, and Brazil. Encouraging results came back—after months in transit—but at a high cost. **Francis Upton** put the expense of the latter expeditions at about $16,000, or, as he figured, $150 per sample returned. Faced with diminishing returns, Edison called off

the project in early 1888. Not all of the costs were monetary. One agent died of yellow fever in Cuba, and another spent months recovering after returning from South America. The expeditions also yielded some detailed travel narratives as well as photographs published by **William K. L. Dickson** and his sister Antonia Dickson. *See also* ELECTRIC LAMP.

FIRESTONE, HARVEY SAMUEL (1868–1938). An American industrialist renowned for manufacturing rubber tires in Akron, Ohio, Firestone became a personal friend of Edison. When he and **Mina Miller Edison**, also from Akron, were young, their families overlapped socially at the **Chautauqua Institution**. Firestone joined the group of self-styled **Vagabonds**, who took yearly automotive camping trips with Edison until 1924. Anxious about the precarious supply of natural rubber from abroad, Firestone and **Henry Ford** encouraged Edison to investigate alternatives that could be cultivated domestically, and they gave financial support to Edison's **rubber research** project.

FORD, HENRY (1863–1947). Famed for pioneering the mass production of inexpensive automobiles, Ford revered Edison and became a personal friend in the twentieth century. In 1891, the young Ford left his small lumber business to work for the Edison Illuminating Company in Detroit, rising quickly to chief engineer. In that role, he met Edison at an 1896 meeting of the Association of Edison Illuminating Companies and received the inventor's encouragement in his automotive dreams. Their next meeting took place in early 1912, when Ford asked Edison to develop a starter battery for Ford automobiles. Ford also agreed to loan Edison $500,000 to build a new **battery (storage)** factory in expectation of large orders from the Ford company. It was at this time that Ford began to give automobiles to Edison and other members of his family. In the late 1920s, **Thomas A. Edison, Incorporated** (under **Charles Edison**'s leadership) developed the Emark lead-acid storage battery for Ford and later sold it to other companies as well.

Beginning in 1914, Ford joined Edison and **John Burroughs** on automotive camping trips that later included **Harvey Firestone** among the **Vagabonds**. During the 1914 trip, they spent time at Edison's winter home in **Fort Myers, Florida**; two years later, Ford purchased the adjacent home originally built for **Ezra Gilliland**. Ford subsequently reconstructed Edison's **laboratory** in **Menlo Park, New Jersey**, at the Henry Ford Museum and Greenfield Village; sponsored a celebration of the fifty-year anniversary of the Edison incandescent light; and became an honorary member of the **Edison Pioneers**. Ford's virulent and very public anti-Semitism has invited retrospective comparison with Edison's own prejudices, which were more common and largely kept from public view. *See also* MENLO PARK LABORATORY IN GREENFIELD VILLAGE.

FORT MYERS, FLORIDA. Edison first visited Fort Myers in 1885, when it was an isolated but relatively affluent cattle entrepôt on the state's central Gulf Coast. Recently widowed, he was traveling with his daughter **Marion Estelle Edison**, friend **Ezra Gilliland**, and Gilliland's wife. Lured by the climate, the surroundings, and the hunting and fishing, Edison and Gilliland bought thirteen acres of land. Each man intended to build his own house (and Edison a **laboratory**) to use as winter residences. Newlyweds Edison and **Mina Miller Edison** honeymooned there in March/April 1886, a period of remarkable imaginative activity for the inventor that included making detailed plans for landscaping the grounds. The couple returned in 1887 as Edison recuperated from illness. Then came the passionate rift with Gilliland in 1888 and a decade of iron-ore milling, and Edison did not again set foot on the property until 1901, leaving its upkeep to caretakers and vacationers. (Gilliland may have sold his acreage in 1891.) In the 1910s and 1920s, Edison regularly made long visits to Seminole Lodge, as the estate had become known, and made it a center of his botanic **rubber research**. Mina Edison deeded it to the city of Fort Myers shortly before her death. **Henry Ford**, a frequent visitor, bought adjacent property in 1916. Today their homes are open to the public and operated jointly as the Edison and Ford Winter Estates. *See also* EDISON STAR.

FRAZAR, EVERETT (1834–1901). An American merchant and diplomat, Frazar provided Edison with ready commercial networks in East Asia, particularly in China, Japan, and Korea. Frazar became the American consul general for Korea in 1884, and his New York trading house (Frazar & Company) had branches in Shanghai, Nagasaki, Yokohama, and Hong Kong. In the mid- to late 1880s, he sought government concessions for Edison's telephone and **electric light and power system** in Korea and represented the electric light and **phonograph** in parts of Asia. A prominent resident of Orange, New Jersey, and an active member of that city's New England Society, Frazar may have had a role in introducing Edison to photographic pioneer **Eadweard Muybridge**. His son Everett Welles Frazar (a Stevens Institute graduate) worked for Edison at the **laboratory** in **West Orange, New Jersey**, from 1890 to 1896.

FUEL CELLS. A fuel cell creates electric current by an electrochemical reaction, somewhat like a conventional battery. However, unlike a battery, the fuel cell is designed for its oxidizable fuel (such as coal) to be replenished as it is consumed. The fuel cell produces free electrons, either directly by the reaction of coal with other substances, such as nitrates, or by the creation of an ionized gas. Edison briefly experimented with a carbon battery of this type in the mid-1870s. With the growth of electric light and power systems in the 1880s, he and several other researchers separately began to explore how such batteries might be used for the **direct conversion** of coal into electricity. Edison sought to generate a current by using heat and an active oxidizing agent to oxidize a carbon electrode or to produce a rarefied ionized gas from carbon or a metal. He also experimented with a finely divided metal and a peroxide (usually manganese) in a solution of sulfuric acid to catalyze the oxidation of the carbon. Edison obtained several patents before shifting his attention around 1887 to other methods of converting fuel into electricity without the usual intermediaries of steam engines and rotating generators.

G

GAS LIGHTING. Systems of gas lighting, first developed in Britain in the 1790s, were common in American and European cities by the late 1870s, and Edison looked to them as a model for electric lighting. The model featured the generation of gas at a central plant and its distribution through large underground pipes to individual buildings, where it was metered as the basis for billing. Smaller pipes took the gas to individual fixtures that could be turned on and off independently. Lamps or burners were standardized to produce a uniform light of about fifteen candlepower (at least in theory). Edison sought to incorporate these familiar elements into his **electric lighting system**, right down to the lamp of approximately fifteen candlepower. Recognizing gas as the entrenched lighting method he would have to displace, he and his staff at the **laboratory** in **Menlo Park, New Jersey**, compared the projected construction and operating costs of electric lighting with those of gas. When planning central stations, they assessed demand in the proposed service area by tabulating its number of gas lights. In its promotional materials, the **Edison Electric Light Company** called out the unpleasant and dangerous effects of gas lighting (heat, fumes, fire, and accidental asphyxiation). To certify the safety of electric lighting, the Edison company sought and received approval from local and national boards of fire insurance underwriters.

Although Edison initially focused on central station lighting, most early incandescent electric lighting came from so-called isolated plants designed to light a single building. Here too, gas light provided a model in the form of small gas plants that became popular in the post–Civil War United States for homes outside of urban gas distribution networks. These systems produced kerosene or gasoline vapor from a tank underground or in an outbuilding and distributed it through interior pipes to individual gas fixtures. Edison had such a system at his Menlo Park laboratory before he had electric lights, and the investors in Edison Electric likely had them in their homes or businesses.

Competition from electric lights led to Carl Auer von Welsbach's incandescent gas mantle, which helped the gas lighting industry remain competitive into the early twentieth century. However, the rapid growth of electric lighting early in the new century led to consolidation between electric and gas utilities. Although gas continued to brighten streets in many cities, it became more commonly used for heating and cooking than for lighting.

GENERAL ELECTRIC COMPANY (GE). The 1892 merger of rivals **Edison General Electric** (EGE) and **Thomson-Houston Electric** symbolized Edison's exit from the electrical industry and created one of the largest American companies at the time. Details of the consolidation are murky, but as Thomson-Houston became more profitable than EGE, **J. P. Morgan** and his associates pushed through a deal on terms not entirely agreeable to EGE president **Henry Villard**. Edison put the best face he could on the arrangement, but having long opposed any concessions to competitors, he could not have been happy. (His rejection of a cooperative agreement with Thomson-Houston a year earlier likely cost EGE a stronger footing in the negotiations toward the eventual merger.) Edison became a director of the new company but showed little interest in the position. Two principal figures associated with him did not join EGE. Villard was passed over as president in favor of Thomson-Houston's **Charles Coffin**, and **Samuel Insull**, who declined a subordinate position, started anew in Chicago. But GE's formation also made Edison a wealthy man, at least on paper. His GE stock shares were worth probably a few million dollars, much of which he sold to finance his venture at the **Ogden ore milling plant** over the next decade.

Edison and GE did not separate entirely. He continued doing research on incandescent lamps, first under terms of a prior contract with EGE and then (from 1895) on retainer from GE. The company, without the word "Edison" in its name, played on its historical connection with the inventor and the iconic lamp at the core of its business. From 1909, it sold its new tungsten-filament lamp (a huge improvement on the carbon filament that remade the market for incandescent bulbs) as the "Edison Mazda Lamp," although it had no connection with Edison himself. When Edison visited GE's research **laboratory** at Schenectady, New York, in 1922—his first visit to the site of his former factory in twenty-five years—newspaper reporters witnessed his reminiscences and reunions with old employees. They also chronicled his meeting with Charles Proteus Steinmetz, the brilliant electrical engineer whom the company was happy to have known as the "Wizard of Schenectady," a nickname that clearly echoed Edison's fame as the **Wizard of Menlo Park**.

GE became a conglomerate and one of the twentieth century's corporate giants. Although still closely associated with electric light and power, it entered a wide range of other fields, including transportation, health care, broadcasting, and financial services. The company sold its lighting business in 2020. *See also* COSTER, CHARLES HENRY; DREXEL, MORGAN & COMPANY.

GILLILAND, EZRA TORRANCE (1848–1903). Edison's longtime friend and intermittent business partner, Ezra ("Ed") Gilliland met Edison when both were telegraph operators in Adrian, Michigan, and again in Cincinnati in the mid-1860s. Both young men inclined to manufacturing and invention, and in 1875, they collaborated on telegraph instruments and the **electric pen**, which Gilliland manufactured. Gilliland also began to help with experiments at Edison's **laboratory** in **Menlo Park, New Jersey**, and in 1878 became an exhibitor for the **Edison Speaking Phonograph Company**. Drifting away from New Jersey, Gilliland landed in **Boston**, where he created the experimental department at the **American Bell Telephone Company** in 1883.

Meeting again at the **Philadelphia International Electrical Exhibition (1884)** in September of that year, the two men resumed their friendship and collaboration. Edison was adrift after the death of his wife **Mary Stilwell Edison**, and he welcomed Gilliland's suggestion to take up long-distance telephony. That project was short lived, but the two soon began developing their own ideas for a system of telegraphing to and from moving trains. Amid a circuitous trip to Florida in 1885, they decided to buy Gulf Coast property for adjacent winter homes and, for Edison, a laboratory. Edison made several visits to Gilliland and his wife Lillian in Massachusetts, where his hosts introduced him to eligible young women, including **Mina Miller**, who would become the inventor's second wife. The two men became exceptionally close during this time, with Edison referring to Gilliland as "Damon," the Greek mythological figure who, with Pythias, personified unshakably loyal friendship.

Leaving American Bell in 1886, Gilliland joined Edison in his laboratory in **New York City**, where he worked on **railway telegraphs**, long-distance telephones, and an improved **phonograph**. With long experience in manufacturing, he was a logical choice to set up the new factory for the **Edison Phonograph Company** late in 1887. Gilliland proposed improvements to the prototype machine, suggestions that may have piqued Edison, coming from someone who perhaps seemed less like Damon than a subordinate in a business relationship. Gilliland also took on the duties of exclusive sales agent for the phonograph.

It was in the role of agent that Gilliland and Edison's attorney **John Tomlinson** suggested that **Jesse Lippincott** buy the North American rights on the phonograph. The three men struck a deal to which Edison, in need of cash, reluctantly agreed in June 1888. Gilliland and Tomlinson each made side agreements with Lippincott—in Gilliland's case, for $250,000 to sell his Edison Phonograph Company agency. The circumstances of these arrangements are unclear, but Edison later became enraged that the money going to Gilliland and Tomlinson might otherwise have been paid to him. He severed all connections with Gilliland and Tomlinson and later sued them (unsuccessfully). He would not reconcile, and he and Gilliland never again overlapped at their joint Florida property. Gilliland returned to Adrian, Michigan, and entered the Gilliland Electric Company with his father and brother. When Edison learned in 1890 that the Gillilands had agreed to manufacture nickel-in-slot phonographs for the **Automatic Phonograph Exhibition Company,** he moved to cut them out of the business. Gilliland eventually relocated to Westchester County, New York, where he built his own laboratory and took out several telephone patents. *See also* FORT MYERS, FLORIDA; NORTH AMERICAN PHONOGRAPH COMPANY.

GILMORE, WILLIAM EDGAR (1863–1928). A trusted administrator and executive in several Edison enterprises, Gilmore entered the Edison orbit in 1881 as an assistant stenographer at the **Edison Electric Light Company** in **New York City.** He then did stenography for **Samuel Insull** until January 1884, when Edison put him in charge of compiling cost estimates for the **Edison (Thomas A.) Construction Department**'s projects. When the Construction Department disbanded in May 1884, Gilmore again became stenographer to Insull, following him to the **Edison Machine Works** at Schenectady, New York, in 1886 and eventually becoming assistant general manager. In 1894, Gilmore took over management of the **Edison Phonograph Works** and the **Edison Manufacturing Company,** the latter of which included the film production division. Within a year, Gilmore eased out **William K. L. Dickson,** whom he suspected of working secretly for a competitor in the film industry. As Edison Manufacturing's vice president, Gilmore successfully negotiated the acquisition of the rival Vitascope Company. After 1900, he convinced Edison to concentrate on film production rather than on licensing patents. He also hired Edwin S. Porter and other pioneering directors to make original films both on location and at the Edison Manufacturing Company's Bronx studio. Gilmore became a stockholder and trustee of the New York Concentrating Works (an Edison iron mining venture) and served as president of the **National Phonograph Company** from 1897 until 1908. He left Edison's employ in 1908 to take the presidency of Essex Press, a printing company.

GLENMONT. Edison bought this twenty-nine-room mansion in **Llewellyn Park** in early 1886, just weeks before marrying **Mina Miller Edison.** It remained the Edisons' home for the rest of their lives. Designed by noted architect Henry Hudson Holly in his signature Queen Anne style, the house was completed in 1882, when owner Henry Pedder began to enlarge it. Pedder enjoyed the property for scarcely two years before he forfeited the entire estate as restitution for embezzlement. Edison bought the house and all its contents, greenhouse, and stables on thirteen and a half acres for $125,000. He and Mina kept the furnishings (from **New York City** decorators Pottier & Stymus, including Hudson River School paintings) largely intact. The domestic staff during their tenure included a cook, a waitress/maid, a laundress, and a nurse/governess; other employees included a groundskeeper, a gardener, and a chauffeur. Edison electrified the house (and several neighboring ones) in 1887 by taking power from his new **laboratory** nearby. He transferred the estate to Mina's name in 1891, presumably to protect against creditors. She sold it to **Thomas A. Edison Incorporated** in 1946 as a memorial to her husband. The company later transferred the property to the National Park Service,

"Glenmont," the home where Thomas and Mina Edison lived after marrying in 1886. *Courtesy of Thomas Edison National Historical Park.*

which operates it as part of the **Thomas Edison National Historical Park**. Both Edison and Mina were reinterred there in 1963. The origin of the name "Glenmont" is uncertain, but it apparently dates from their first years there.

GOLD AND STOCK TELEGRAPH COMPANY. This company was organized in 1867 to exploit the small printing telegraph devised by inventor Edward Calahan. Calahan's device, which became known as a **stock ticker**, initially served only the New York Stock Exchange. The company soon expanded into other financial markets and in 1869 acquired its principal rival, Samuel Laws's Gold Reporting Telegraph. Edison, who was superintendent of the Laws company when it was acquired, was soon under contract to improve Gold and Stock's printing telegraph technology. He developed a close relationship with company president **Marshall Lefferts**, who mentored the

young inventor. In 1871, Edison produced his first significant invention, the Universal stock printer, which became the standard Gold and Stock ticker. That May, Edison signed a five-year contract with Gold and Stock as a consulting electrician at an annual salary of $2,000; he also received $35,000 of its stock for rights to his printing telegraph inventions. At the same time, **Western Union** acquired Gold and Stock as a subsidiary, a merger that marked the start of Edison's relationship with Western Union.

GOULD, JAY (1836–1892). Gould was an American financier and speculator whose forays into the telegraph industry in the 1870s helped advance Edison's career as an inventor. Gould was widely known for unscrupulous business practices, including an 1869 scheme to corner the gold market that caused financial panic and ruined many investors. Yet he emerged from that scandal with the wherewithal to buy up railroads, including the Union Pacific. He amalgamated the telegraph companies connected to his railroads and purchased other telegraph properties, including the **Atlantic and Pacific Telegraph Company** (A&P), all in an effort to compete with the **Western Union Telegraph Company**. Gould also purchased the rights to Edison's **automatic telegraph** and **quadruplex telegraph** systems, the latter in a controversial deal that brought Edison into conflict with **William Orton**, his patron at Western Union. Gould further antagonized Orton by hiring Edison as chief electrician of A&P in late 1874. Edison installed his automatic system on the company's lines and developed a district telegraph but largely stopped active work for the company within a year. However, Gould persisted and by cutthroat competition was able to weaken Western Union's market position enough to force an 1881 merger by which he took control of the giant company. *See also* ECKERT, THOMAS THOMPSON.

GOURAUD, GEORGE EDWARD (1842–1912).

An American based in London, Gouraud served for many years as Edison's principal business agent in Great Britain and the empire. He also had roles in selling or licensing Edison patents in some European countries, their colonies, and parts of Asia and the Americas. He was uniformly known as "Colonel," reflecting the rank he attained in the Union cavalry in the Civil War. After the war, he became associated with **George Harrington** and went to London on behalf of various interests. He met Edison in 1873, when the latter attempted to introduce his **automatic telegraph** there. For two decades thereafter, Edison entrusted Gouraud with commercializing a succession of inventions, including the tinfoil phonograph, telephone, electric light, and wax-cylinder phonograph. Like **Edward Johnson**, Gouraud had a flair for promotion. His critics in Edison's inner circle (notably **Samuel Insull**, who had been Gouraud's secretary) pointed to his self-promotion and the gaps between what he promised and what he managed to do. Although the companies Gouraud

formed were not financially successful, they brought the phonograph, telephone, and electric light to limited but influential publics in the financially important British market. As part of his marketing campaign, Gouraud made some significant early recordings in the late 1880s with the new wax-cylinder phonograph, including by poets Alfred Lord Tennyson and Robert Browning. However, his slow pace and equivocation about shouldering significant financial risk himself frustrated Edison, who eventually decided to bypass his old friend by shifting control of the phonograph to a new syndicate. Although Gouraud's behavior grew increasingly eccentric in the 1890s, he had the means to pursue his own inventive interests. He set up an experimental **laboratory** at Hove, England, and financed the sound amplification experiments of inventor Horace Short, resulting in the creation of the Gouraudphone. He also financed aerial balloon experiments.

GRAMOPHONE. In the late 1880s, **Emile Berliner**, a German émigré (and sometime rival of Edison) created this sound recording and playback machine in Washington, D.C. The gramophone differed from both the **phonograph** and the **graphophone** in two major respects. Its recording surface was a disc (not a cylinder), across the flat face of which the recording was inscribed in a spiral groove. Both Edison and **Alexander Graham Bell** had experimented with flat plates, but Berliner managed to overcome the distortion caused by the variation of groove speed with the change in the spiral radius. The other difference (less obvious visually) was its method of incising sideways (not up and down) in the groove. Lateral recording produced sounds that were both higher in quality and louder on playback than cylinders. Berliner also found that lateral recordings on a flat surface could be duplicated with relative ease, giving his machine a significant commercial advantage over cylinder recordings. Berliner did not produce a motor-powered version of his machine until 1896, after which sales of the machine and recordings soared. The **Victor Talking Machine Company** (established in 1901) was based on this invention.

GRAPHOPHONE. The graphophone was a sound recording and playback machine created around 1884 by Chichester Bell and Charles Tainter, associates of **Alexander Graham Bell**, at Bell's laboratory in Washington, D.C. It resulted from "Alec" Bell's 1881 decision to explore sound recording, a field that Edison had largely abandoned. It had obvious similarities to Edison's **phonograph**, but C. Bell and Tainter achieved better sound quality by using a wax compound (rather than tinfoil) as a recording medium. Another fine but legally significant difference was that the recording needle cut away or incised the recording material rather than simply deforming it as Edison machines did. Preliminary merger talks in 1885 between the Bell interests and the **Edison Speaking Phonograph Company** went nowhere. Subsequent publicity around the machine and the creation of a company (by **Gardiner Hubbard**) to manufacture and market it for business dictation (including the possibility of sending recorded cylinders through the mail) prompted Edison to return to developing his phonograph in 1887. He at first instructed his representatives to have nothing to do with the "phonograph pronounced backward graphophone" or its promoters— "pirates," he called them (*TAEB* 8:768, 9:287 n. 3). A year later, however, he was persuaded to allow **Jesse Lippincott** to combine the sales rights for the graphophone and phonograph in the new **North American Phonograph Company**. Subsequently, when **George Gouraud** was stalemated in his efforts to market the phonograph abroad, Edison agreed as well to a unification of foreign interests under the Edison United Phonograph Company. The **Edison Phonograph Works** took over the graphophone factory in Hartford, Connecticut, as part of the latter deal. Although the graphophone clearly vexed Edison, it had only limited commercial success.

GRAY, ELISHA (1835–1901). *See* ACOUSTIC TELEGRAPH; AMERICAN BELL TELEPHONE COMPANY; WESTERN ELECTRIC MANUFACTURING COMPANY.

GRIFFIN, STOCKTON LEE (1841–1921). A native of Delaware, Ohio, Stockton Griffin served during the Civil War as a musician in the 14th Illinois Infantry. Following the war, he worked as a telegrapher, including an 1867 stint with Edison in Cincinnati. Griffin also toiled for the **Gold and Stock Telegraph Company** and for several years managed the eastern wires in **Western Union**'s main **New York City** office. In 1878, he resigned his position at Western Union to become Edison's personal secretary. The hiring was probably precipitated by the large amount of correspondence and relentless stream of visitors Edison received after inventing the **phonograph**. Griffin adopted a collegial, even fraternal tone toward his new employer (the two were born roughly five years apart), addressing Edison in correspondence as "Dear Al," "Friend Al," or "Friend Tom." Griffin resigned in February 1881 for reasons that remain unclear. He eventually moved to California and resumed working as a telegrapher. Griffin died in Napa in 1921.

H

HAMMER, WILLIAM JOSEPH (1858–1934). An electrician and engineer, Hammer played a crucial part in presenting Edison's electric light system to the public in the United States and abroad. He joined the staff of the laboratory in **Menlo Park, New Jersey**, in late 1879 after a year working with **Edward Weston** in **Newark, New Jersey**. Hammer conducted lamp tests until Edison named him chief electrician at the lamp factory at Menlo Park. He supervised the shipment of an Edison "Jumbo" steam dynamo to the **Paris International Electrical Exposition (1881)**, later following the machine to Paris to install and operate it as part of Edison's exhibit—the first of many public Edison lighting displays made by Hammer or with his assistance. From Paris, he went to London to help **Edward Johnson** construct the first Edison central station, a demonstration plant at **Holborn Viaduct** (of which he became chief engineer). He helped put together the Edison exhibit for the **London Crystal Palace International Electrical Exhibition** of 1882, for which he designed an electric sign spelling out Edison's name in incandescent lights—the first such motor-driven flashing sign. He made a similar flashing sign for an 1883 exhibition in Berlin. Returning to the United States, he took charge of Edison exhibits at the **Philadelphia International Electrical Exhibition (1884)** and the Cincinnati Electrical Exhibition (1888). Hammer's tour de force came in 1889, when he oversaw the massive Edison display at the **Paris Universal Exposition**. Fifteen years later, he received the grand prize at the St. Louis International Exposition for his exhibit of incandescent lamps collected over more than three decades.

Hammer's other roles in the Edison lighting industry included chief engineer of **Deutsche Edison Gesellschaft** (1884), chief inspector of central stations for the **Edison Electric Light Company** (1884–1885), and chief engineer and general manager of the Edison illuminating company in **Boston** (1886–1887). Hammer also received credit for first using radium on the hands of watches, clocks, and other instruments and for suggesting the use of radium to treat cancer. For contributions to military engineering, he was named a major in the U.S. Army and, in 1925, a chevalier of the French Legion of Honor. Hammer helped establish the **Edison Pioneers** and served as its president in 1920.

HANDWRITING. Edison used three distinctive styles of handwriting, each for a different purpose. He first developed a calligraphic hand, one that he later reserved for special occasions and formal letters to important figures. For writing rapidly to himself or close associates or for making notebook entries, marginal notes, and draft patents, he used a loose and informal script. And for everyday correspondence, he used a neat but informal hand. Edison developed this quotidian handwriting as a young telegraph operator rapidly transcribing press reports. He formed each letter separately without any connecting lines, making for a highly legible style and allowing him to take down some forty-five words per minute with less fatigue than by any other method. In 1869,

FROM THE LABORATORY OF
T. A. EDISON,
MENLO PARK, N. J.
U. S A.

To the Editor of the Daily Graphic   May 10 1878.

Dear Sir - I feel an inclination to thank you for the pleasant things you have said about me and the Phonograph in the Graphic. Your words and pictures have gratified me the more, because I had long since come to look upon your paper with pride and to regard such an illustrator of daily events, as one of the marvels of the age. I am able to report to you that I am constantly increasing the sensibility and power of the Phonograph. I feel certain that it will soon justify all the hopes of its friends. By the way. Croffuts April-first hoax concerning my alleged food machine has brought in a flood of letters from all parts of the country. It was very ingenious. With congratulations on the great success of your journal I am

Yours Truly

Thomas. A. Edison.

An example of Edison's calligraphic script. His letter to the editor of the *Daily Graphic*, a New York newspaper, mentions an April Fool's Day hoax about his invention of a machine to create food. *Courtesy of Thomas Edison National Historical Park.*

the trade paper *The Telegrapher* remarked that by adopting this telegraph script, Edison had made himself "about the *finest* writer we know of" (*TAEB* 1:75). When *Science* reported on his script in 1885, it came to the attention of librarians C. Alex Nelson and Melvil Dewey, who were seeking the most efficient and legible writing for library catalog cards. Within two years, libraries across the nation had adopted Edison's hand. While tests showed note-taking hand to be quicker, the Edisonian script, known today as library hand, was more legible, and librarians made fewer mistakes with it.

HARRINGTON, GEORGE (1816–1892). A business partner with Edison in **automatic telegraphy**, Harrington began his public service career as a Navy Department clerk during the Tyler administration. He then clerked in the Treasury Department, rising to assistant sec-

retary under Abraham Lincoln—whose funeral he organized. He then served as U.S. minister to Switzerland, where he could observe government telegraph services across Europe. After returning to the United States in 1869, he became interested in the commercial possibilities of automatic telegraphy. Harrington helped to support Edison's work in that area as a partner in the **American Telegraph Works** and as president and investor of the **Automatic Telegraph Company.**

HELMHOLTZ, HERMANN VON (1821–1894). Helmholtz was among the foremost German scientists of his day. Trained as a physician, his expertise encompassed physiology, chemistry, and physics, all of which informed his investigations of the physiology of sensation during his tenure at Heidelberg University (1858–1871). The science of acoustics figured prominently in his work from the perspectives of both physiology and pure physics. He published a masterwork (*On the Sensations of Tone as a Physiological Basis for the Theory of Music*) in 1862 that Edison studied when it appeared in English in 1875. In it, Helmholtz described "resonators," hollow vessels of glass or metal tuned so as to detect or amplify specific frequencies. Edison constructed such resonators for his 1875 experiments in **acoustic telegraphy**. He used them again three years later in his "auriphone," a hearing-aid device that did not live up to its initial promise. Helmholtz had indirect influence on Edison's **laboratories** through the training of two assistants: **Francis Upton** (who studied for a year with him in Berlin) and Franz Schulze-Berge (a chemical expert at **West Orange, New Jersey**, whose doctoral research he codirected).

Helmholtz put forward a theory to explain the formation of vowel sounds, a contested phenomenon. In essence, he posited that a fundamental, unique tone produces each vowel's distinctive tone. Edison hazarded a criticism, but it fell to two British scientists (Fleeming Jenkin and J. A. Ewing) to disprove it by using a **phonograph** to change the pitch of a vowel sound without altering its essential character.

In 1889, Helmholtz and **Werner von Siemens** hosted a dinner in Berlin to honor Edison. The three then traveled to a scientific conference in Heidelberg. On returning to the United States, Edison sent several phonographs to Helmholtz—one for Siemens and two for the Gesellschaft Urania, a public observatory and science center in Berlin. Edison later recalled regaling Siemens with humorous stories on the Heidelberg trip, but Helmholtz (who knew English) was unmoved. And at some later date, Edison inscribed the flyleaf of his personal copy of Helmholtz's 1875 English edition (now at the **Thomas Edison National Historical Park**) with the note, "Helmholtz as I can personally vouch had absolutely no sense of humor." And although he turned repeatedly to the book throughout his career, he remarked on the same page that it was "an immense mass of things stated in the most muddy man-

ner. A masterpiece of incapacity of explaining simple things." Helmholtz and Edison met on two other occasions: at the 1893 **World's Columbian Exposition** in Chicago and three years later when Columbia University hosted a reception for Helmholtz.

HENRY, JOSEPH (1797–1878). Born in Albany, New York, to Scottish immigrants, physicist Joseph Henry taught science and mathematics at the Albany Academy and the College of New Jersey (later Princeton University) before becoming the first director of the Smithsonian Institution in 1846, a position he held until his death. An expert in electromagnetism and its effects, Henry discovered the principle of self-induction and invented both the electromagnetic relay and a reciprocating electric motor. During his Smithsonian tenure, he also set up a network of weather observation stations, laying the foundation for the U.S. Weather Bureau. Henry took notice of the young Edison, reportedly calling him "the most ingenious inventor in this country . . . or in any other" (*TAEB* 3:628). As president of the **National Academy of Sciences** (1868–1878), he invited Edison to exhibit his **phonograph** and **carbon-button telephone** at the Academy's April 1878 meeting in Washington, D.C. On 18 April, Edison and **Charles Batchelor** demonstrated the phonograph and carbon telephone at the Smithsonian. The presentation astonished those present, and the crowd was such that the doors had to be taken from their hinges to accommodate more spectators. Henry died less than a month later and was buried in Washington, D.C. A bronze statue of him stands in the Main Reading Room of the Library of Congress's Thomas Jefferson Building. In 1916, Edison served on an honorary committee to raise funds for a memorial statue in Albany.

HIGH RESISTANCE. The principle of high resistance was key to the practical success of Edison's incandescent **electric lamp** and his **electric light and power system** generally. The idea was, as he put it, "the more resistance your lamp offers to the passage of the [electrical] current, the more light you can obtain with a given current" (Friedel, Israel, and Finn 2010, 58). At a time when few comprehended Ohm's law (an algebraic expression of the proportional relations among current, resistance, and voltage in an electric circuit), Edison instinctively grasped its importance almost from the start. Electric lighting in 1878 meant arc lamps, which had a resistance of one or two ohms; Edison was trying to get 100 or more times that figure to bring a "burner" or filament to glowing heat with a modest current. The usefulness of resistance was misunderstood by physicists and engineers accustomed to thinking of it as wasteful in arc lighting, electroplating, and other familiar work. They assumed that Edison would need rivers of energy—and ruinously expensive generators and conducting wires—to run hundreds of lamps from a single source in parallel ("multiple arc") circuits. Edison's intuitions were validated by calculation and experiment. He designed, built, and operated his first central station systems around lamps with about 140 ohms of resistance. In the long term, Edison never stopped trying to increase efficiency through higher resistance. When his basic lamp patent was challenged in court, it was upheld not on the basis of the carbon filament but largely on the originality of the principle of high resistance. *See also* SAWYER, WILLIAM EDWARD.

HOLBORN VIADUCT. This bridge, completed in 1869 along a quarter mile of a commercial corridor in the City of London, was the site of Edison's first full-scale electric central station. In 1881, **Edward Johnson** and banker Egisto Fabbri chose 57 Holborn Viaduct for a temporary showcase of Edison's new **electric lighting system**, and Johnson and **William Hammer** oversaw construction. When the station opened in April 1882, two large ("Jumbo") dynamos powered 164 street lights, 542 lamps in about two dozen shops and public buildings along the Viaduct, and another 232 lamps in the station itself. Because the station was a demonstration rather than a commercial enterprise, it initially gave away its power for free and later charged customers no more than

what they had been paying for **gas lighting**. The station was operated by the Edison Electric Light Company Ltd. (the Edison company in Britain) and its successor until 1886, with a brief hiatus in 1884.

HONORS AND AWARDS. Major honors and awards granted to Edison in his lifetime.

EXHIBITION AWARDS

1870: American Institute Fair (**New York City**): First prize for "Best Electric Telegraph Instrument" (with **Franklin Pope**).

1876: **Centennial Exhibition** (Philadelphia): Prize medals for the automatic telegraph system and electric pen and autograph press.

1878: **Paris Universal Exposition**: Grand Prize for the phonograph, telephone, and electric pen and autograph press.

1878: American Institute of the City of New York: Medals for **electric pen** and multiplying press.

1879: American Institute of the City of New York: Medals for chemical telephone (**electromotograph receiver**) and **carbon-button telephone transmitter**.

1881: **Paris International Electrical Exposition**: Diploma of the first class for electric lighting.

1882: **London Crystal Palace Exhibition**: Gold medal.

1883: Southern Exposition (Louisville, Kentucky): Four first-place awards for electric lighting.

1889: **Paris Universal Exposition**: Diplome d'Honneur; Grand Prix for electrical appliances exhibit.

1915: Panama Pacific International Exhibition: Medal of Award.

MEDALS AND HONORARY TITLES

1878: Chevalier of the French Legion of Honor.

1881: Officer of the French Legion of Honor.

1889: Commander of the French Legion of Honor; Grand Officer of the Crown of Italy.

1892: Albert Medal (British Society of Arts).

1895: Rumford Prize (American Academy of Arts and Sciences).

1899: Edward Longstreth Medal (Franklin Institute).

1908: John Fritz Medal (American Association of Engineering Societies).

1909: Edison Medal [inaugural award] (American Institute of Electrical Engineers).

1915: Franklin Medal in Engineering (Franklin Institute).

1920: Distinguished Service Medal (U.S. Navy).

1928: Congressional Gold Medal (U.S. House of Representatives).

1928: Gold Medal (Society of Arts and Sciences).

1929: Honorary Academy Award (Academy of Motion Picture Arts and Sciences).

HONORARY DEGREES AND MEMBERSHIPS

1878: Fellow, **American Association for the Advancement of Science.**

1878: Doctor of Philosophy, Union College (Schenectady, New York).

1879: Doctor of Philosophy, Rutgers College (New Brunswick, New Jersey).

1890: Berlin Elektrotechnischer Verein.

1915: Doctor of Science, Princeton University.

1917: Doctor of Laws, University of Pennsylvania.

1922: Doctor of Science, Rutgers College.

1927: Member, **National Academy of Sciences**.

1930: Doctor of Science, Rollins College (Winter Park, Florida).

HOPKINSON, JOHN (1849–1898). A distinguished electrical engineer and physicist, Hopkinson was a consulting engineer for the Edison Electric Light Company Ltd in London from 1881 to 1883. His work there overlapped with Edison's own in the design of electrical distribution systems and **generators**. In July 1882, Hopkinson filed a provisional patent specification for a **three-wire system** of electrical distribution that greatly reduced the amount of expensive copper conductors in central station power networks. Edison had the germ of this idea at about the same time, though he did not file for a U.S. patent until November of that year. Edison apparently knew nothing of Hopkinson's work until early 1883, and their nearly simultaneous development of the concept is generally seen as a case of independent coinvention.

Hopkinson used his mathematical gifts to make pioneering studies of generator performance, and he set out to increase the electrical output of Edison's machines with respect to physical size and weight. He made a number of alterations between mid-1882 and early 1883, the most visible being to Edison's distinctively long and slender field magnets. Making them shorter and thicker, he created a more efficient magnetic circuit. Edison learned of Hopkinson's efforts in late 1882, just as he was making his own improvements to his dynamo. There was no direct communication between the two men, but at key moments in early 1883, **Edward Johnson** and **Charles Batchelor** informed Edison of Hopkinson's progress. By that summer, Edison and Hopkinson had produced machines of similar proportions. Although their windings and electrical characteristics differed, the machines were typed as "short-core" dynamos and widely referenced in the United States as the "Edison–Hopkinson" style. Edison adopted the shorter field magnets in all subsequent dynamo models.

HOUSTON, EDWIN JAMES (1847–1914). *See* EDISON EFFECT; THOMSON-HOUSTON ELECTRIC COMPANY.

HUBBARD, GARDINER GREENE (1822–1897). Hubbard was a Boston attorney best known as the father-in-law of **Alexander Graham Bell** and the founder and first president of the **Bell Telephone Company**. Hubbard was active in local Boston institutions; in 1867, he helped establish the Clarke School for the Deaf, the first of its kind in the United States. By that time, he had risen to national prominence as a leading critic of the **Western Union Telegraph Company** and a proponent of nationalizing the telegraph system. His efforts to counter Western Union's monopoly power led to him to support Bell's experimental work on a system of **acoustic telegraphy.** (Bell was teaching Hubbard's deaf daughter Mabel, whom he would subsequently marry.) When Bell's experiments led to the creation of the telephone, Hubbard played a key role in organizing the Bell Telephone Company to market the device. In 1878, he became an organizer and principal investor in the **Edison Speaking Phonograph Company**, formed on behalf of Edison's new invention. During the 1880s, Hubbard supported the efforts of Bell and his Volta Laboratory associates to develop and promote the **graphophone**, their version of the wax-recording phonograph. Hubbard was also a founder and the first president of the National Geographic Society. *See also* AMERICAN BELL TELEPHONE COMPANY.

HUTCHISON, MILLER REESE (1876–1944). Hutchison played a major role in promoting the Edison **storage battery** business in the crucial years before and during World War I. A native of Montrose, Alabama, he studied electrical and mechanical engineering at what is now Auburn University. He received his first patent—one of more than 100 in his career—in 1895. He served as an electrical engineer in the Spanish-American War and then moved to **New York City**. Hutchison had several profitable inventions to his credit (including the Klaxon automobile horn and Acousticon hearing aids) when he became a consultant on the Edison storage battery in 1910. Edison gave him the title of chief engineer of the **laboratory** in **West Orange, New Jersey**, two years later, though Hutchison worked mainly on the battery and spent much of his time promoting it to federal officials in Washington, D.C. It was Hutchison who approached Navy Secretary **Josephus Daniels** about creating an advisory body that became the **Naval Consulting Board**. Despite working for Edison and the **Edison Storage Battery Company** (ESBC), Hutchison devised a middleman role for himself that generated lucrative commissions on battery sales. His self-serving business methods soured relations with ESBC and principals of **Thomas A. Edison Incorporated**, notably **Charles Edison** and **Stephen Mambert**. After a navy inquiry faulted a defective Edison battery for a deadly blast aboard the submarine *E-2* in January 1916, Hutchison's overzealous disputation of the findings further damaged his standing. He resigned from ESBC in June 1918 and from the laboratory a month later, cutting all ties with Edison. Hutchison set up his own laboratory and continued to produce inventions, including a "super cannon" to shoot a projectile 300 miles.

I

INSULL, SAMUEL (1859–1938). One of Edison's closest associates for eleven years, Samuel Insull was born in London to a clergyman and the keeper of a temperance hotel. **George Gouraud** hired him as secretary in 1879, and he caught the eye of **Edward Johnson**. On Johnson's recommendation, Insull became Edison's private secretary in February 1881, succeeding **Stockton Griffin**. Insull quickly earned Edison's trust and became his de facto business manager, assuming general oversight of Edison's financial affairs and coordinating the work and financing of the **Edison Construction Department** and the electrical manufacturing shops. He clearly relished the personal authority that came with these roles and was able to keep pace with Edison's legendary working hours. Insull oversaw the move of the **Edison Machine Works** to Schenectady, New York, and served as its secretary, treasurer, and general manager. He ceded his role as Edison's personal secretary to **Alfred Tate** in 1887. He was also a stockholder, director, and executive in other Edison businesses, including a vice president of the **Edison General Electric Company** (EGE). Insull streamlined EGE's functional organization and had effective responsibility for its large manufacturing operations. Details of his role in the 1892 consolidation of EGE with **Thomson-Houston Electric** into the **General Electric Company** (GE) are unclear, but the merger led to what he later called "the only misunderstanding with Mr. Edison of the entire period of my connection with him" (Israel 1998, 336).

Insull resigned from GE in 1892 to lead the Chicago Edison Company (later Commonwealth Edison). In Chicago, he built up a multi-billion-dollar utility empire across thirty-seven states that reportedly generated one-tenth of the nation's electricity in the late 1920s. However, the Great Depression led to the collapse of his various holding companies, causing catastrophic losses for thousands of investors. Accused of fraud and embezzlement, Insull fled to Europe but was extradited to Chicago for trial. After three acquittals, he returned to Europe and died in Paris.

J

JEHL, FRANCIS (1860–1941). Jehl was a relatively unskilled but important assistant at the **laboratory** in **Menlo Park, New Jersey,** who contributed to Edison's electric light research and later published a memoir of his time there. A native New Yorker, Jehl had worked for **Grosvenor Lowrey** and then **Western Union** while studying at the Cooper Union night school. Lowrey recommended him to Edison in early 1879 as having "a rather awkward appearance, and manners, and is rather slow and might seem to some stupid, [but] he is quite an intelligent, industrious, faithful, honest, high minded young fellow" (*TAEB* 5:86). Jehl did not disappoint, performing mundane but necessary tasks such as maintaining batteries, operating mercury vacuum pumps, and testing meters. He also helped make significant design changes to the vacuum pumps used in the laboratory and Edison's lamp factory. He kept a diary of activity at the laboratory through the early part of 1881, until about the time he took charge of the testing room at the **Edison Machine Works.** He went to Europe in 1882 to help introduce the Edison light system and remained abroad until 1922, spending much of that time as chief engineer of the central station in Budapest. After Jehl returned to the United States, **Henry Ford** hired him to supervise reconstruction of the Menlo Park laboratory at the Henry Ford Museum and Greenfield Village. Late in life, Jehl published *Menlo Park Reminiscences,* a colorful and historically useful—if somewhat unreliable—account of work and events at

the laboratory. *See also* EDISON PIONEERS; VACUUM TECHNOLOGY.

JOHNSON, EDWARD HIBBERD (1846–1917). As an electrician, executive, promoter, and friend, Johnson played crucial roles in commercializing Edison's inventions in the 1870s and 1880s. A former telegrapher, he met Edison as superintendent of the **Automatic Telegraph Company** in the early 1870s. He periodically assisted with experiments at Edison's **laboratories** in **Newark, New Jersey,** and **Menlo Park, New Jersey,** and in 1876 became an officer of the American Novelty Company. For more than a dozen years thereafter, Johnson was an able administrator and tireless booster for the **phonograph,** telephone, and **electric light system.** In two extended stays in Great Britain, he acted as Edison's liaison to the scientific and business establishments. Johnson held technical or managerial positions in the **Edison Speaking Phonograph Company,** the Edison Telephone Company of London Ltd, and the Edison electric light company in Britain. He was a partner in the **Edison Lamp Company** and **Bergmann & Company** and served as president of the **Edison Company for Isolated Lighting** and the **Edison Electric Light Company.** It would be difficult to overstate his loyal efforts, which a London newspaper summed up this way in 1882: "There is but one Edison, and Johnson is his prophet" (*TAEB* 6:312). Relations between the two men were strained in the late 1880s by divergent interests in the formation of **Edison**

General Electric Company and the Edison Phonograph Company and by Johnson's business connections with Frank Sprague. Afterward, Johnson and Edison remained friendly but never again had such close personal or business ties. Johnson is widely credited as the first person to decorate a Christmas tree with incandescent lights (1882).

JORDAN (SARAH) BOARDINGHOUSE. The Jordan boardinghouse in Menlo Park, New Jersey, served as home for a number of Edison's unmarried employees. With Edison's encouragement, Sarah Jordan (1837?–1910), a widow with a young daughter and a distant relative of Mary Stilwell Edison, moved to Menlo Park in late 1878 and opened the boardinghouse. The wood frame duplex was a short walk from and in full view of Edison's laboratory. Jordan and her daughter lived in one half, with ten to sixteen of Edison's men in the other half at any one time. Francis Upton, Francis Jehl, and Stockton Griffin lived there, while others (such as Charles Clarke) only took meals there. Jordan's boardinghouse was among the buildings lighted for the first public display of Edison's electric light at the end of 1879 (at least one other Menlo Park rooming house was electrified in 1880). In 1928, the house structure was dismantled and moved with the laboratory buildings to the Henry Ford Museum and Greenfield Village in Dearborn, Michigan.

K

KENNELLY, ARTHUR EDWIN (1861–1939). Kennelly played an indispensable role as Edison's lead electrical assistant at his **laboratory** in **West Orange, New Jersey**, in its early years but achieved wider renown after leaving. Born in India to English parents, Kennelly was educated mostly in England until age fourteen, when he became an office boy and assistant secretary to the Society of Telegraph Engineers in London. He hired on with the Eastern Telegraph Company as a clerk and rose through the ranks, acquiring a working knowledge of electricity and the title of senior chief electrician—all while publishing several articles. While in the United States in September 1887, he asked **Ezra Gilliland** about working for Edison. By December, he was head of the laboratory's Galvanometer Room. There he helped theorize and test a long list of electrical technologies, including **batteries, electric generators**, meters, **electric motors**, and transformers. Kennelly also helped with experiments or planning related to **electromagnetism**, hydrogeneration at Niagara Falls, medical applications of electricity, and **electrocution**. He left West Orange in 1894 to form a partnership with Edwin Houston as consulting engineers, though in 1901 he presented Edison's paper on the **alkaline storage battery** to the **American Institute of Electrical Engineers**. Finding that existing theory could not explain the transmission of radio waves across the Atlantic, Kennelly in 1902 inferred the existence of a reflective layer of ionized gas in the upper atmosphere. This discovery, made independently at the same time by Oliver Heaviside, was named the

Kennelly–Heaviside layer. Kennelly served as a professor of electrical engineering at Harvard (1902–1930) and of electrical communication at the Massachusetts Institute of Technology (1913–1925). He authored Edison's posthumous biographical memoir for the **National Academy of Sciences**.

KINETOPHONE. From the beginning of his work on **motion pictures**, Edison thought about combining moving images and the **phonograph** to produce films with sound. In 1894–1895, he and **William K. L. Dickson** devised a system for mechanically synchronizing the **motion picture camera** with a phonograph. The combination allowed for the exhibition of films with a peephole **kinetoscope** and a phonograph that reproduced the accompanying sound through earphones. Dickson called the system a "kineto-phonograph" in an 1894 *Century Magazine* article, but Edison referred to it as the kinetophone. Although competitors introduced other sound-film systems over the next ten years, Edison did little with his until 1907. After several years of work, he created a new system that synchronized cameras and projectors with large-cylinder phonographs to record and play back not only music but voices as well. The new system could also project both image and sound for theater audiences. The Motion Picture Division of **Thomas A. Edison Incorporated** began producing kinetophone films in 1912 and soon distributed the system to theaters in the United States, Canada, and Europe. However, by the end of 1914, the war in Europe and a disastrous fire at

Edison's factories and offices in **West Orange, New Jersey**, led to the kinetophone's demise. One early film that survived—in which Dickson plays the violin while two men dance—has been synchronized using modern technology, as have several later examples.

KINETOSCOPE. The original Edison device for exhibiting **motion pictures**, the kinetoscope was a wooden box with a peephole eyepiece through which an individual viewer could watch images produced by a moving strip of film. **William K. L. Dickson** developed it, and Edison patented the kinetoscope in 1891 (U.S. Patent 493,426), about the time he demonstrated it at his **laboratory** to members of the National Federation of Women's Clubs. Edison publicly showed an improved version at the Brooklyn Institute of Arts and Sciences in 1893, and the first commercial kinetoscope parlors opened the next year. However, the kinetoscope was a short-lived phenomenon, as projected film soon became the standard form of exhibition. In 1896, Edison acquired rights to a projector that he marketed as the Edison Vitascope.

KRUESI, JOHN (1843–1899). Swiss-born machinist and mechanical engineer John Kruesi was one of Edison's longest-tenured associates. Kruesi received his early training at a machine shop in Switzerland. He came to Paris in 1867 and worked in several industries over the next three years. At the outbreak of the Franco-Prussian War, he moved to London and then to the United States. Following employment at the Singer Sewing Machine Company in Elizabeth, New Jersey, Kruesi worked for the Edison & Unger Company in **Newark, New Jersey**, which made Edison's patented stock tickers and telegraph instruments. When Edison opened his experimental shop in Newark, Kruesi went with him. He then followed Edison to the new **laboratory** in **Menlo Park, New Jersey**, where he made special tools and precision models and contributed some design work. (It was Kruesi who built the first phonograph from Edison's sketch in 1877.) He was said to be one of the most indefatigable of Edison's employees.

In 1881, Kruesi became treasurer and general manager of the new **Electric Tube Company**, which manufactured underground conduits for Edison's lighting system. He oversaw installation of conductors for the **Pearl Street Station** ("First District") in **New York City** (1881–1882) and for the **Edison Construction Department** (1883–1884). He also received several patents on tube designs. When the **Edison Machine Works** (EMW) absorbed the Electric Tube Company, Kruesi became its assistant general manager. He moved with EMW to Schenectady, New York, in 1886 and remained in that position when the **Edison General Electric Company** (EGE) in turn subsumed EMW in 1889. He supervised the design and construction of new shops at Schenectady between 1886 and 1892 as the number of employees rose from 200 to more than 4,000. When EGE consolidated with the **Thomson-Houston Electric Company** to form the **General Electric Company** (GE) in 1892, Kruesi became the new company's general manager. In 1896, he was appointed GE's chief mechanical engineer, overseeing its factories in Schenectady and New York City until his death.

L

LABORATORIES. In the 1870s, Edison began to merge traditional mechanical invention (and its machine shop practices) with sophisticated electrical and chemical laboratories. In the process, he helped to invent the modern industrial research laboratory. Edison set up his first laboratory in a corner of his telegraph manufacturing shop in **Newark, New Jersey**, in the fall of 1873, prompted by his encounter with the more advanced British electrical community. The fine test instruments in Britain enabled practitioners to make precise measurements and solve problems arising from undersea and underground telegraph lines (common in Britain) that had proven troublesome for Edison's tests of his **automatic telegraph** system there. Within six months of returning from England, Edison could boast that his new laboratory contained "every conceivable variety of electrical apparatus, and any quantity of chemicals for experimentation" (*TAEB* 2:104). In his new laboratory, Edison began to focus his experiments on applied research into electrical and electrochemical phenomena rather than on the electromechanical designs that had made his reputation as a telegraph inventor. He was joined there by a new experimental assistant, **Charles Batchelor**.

By the spring of 1875, Edison decided to make his laboratory entirely independent of the manufacturing shop. He retained some of the machine tools and machinists for a small experimental shop within his growing lab, which now encompassed two floors of the shop building; the other half remained a manufacturing enterprise operated by his former partner **Joseph Murray**. Edison decided in late 1875 to further expand his facilities by building his now famous laboratory in **Menlo Park, New Jersey**.

The Menlo Park facility anticipated a new approach to invention, as Edison merged the shop tradition with increasingly sophisticated chemical and electrical laboratory research. When opened in April 1876, it was probably the best-equipped private laboratory in the United States and certainly the largest devoted to invention. Initially funded by Edison himself, the laboratory also had support over the years from **Western Union**, the **Edison Speaking Phonograph Company**, and the **Edison Electric Light Company**. These resources allowed Edison to increase his staff from the two experimenters and three machinists who moved with him from Newark to more than fifty. Among them were college-educated scientists and engineers, self-taught experimenters and assistants, skilled machinists and carpenters, and office workers to deal with correspondence, purchases, and cost accounting. Money from the Edison Electric Light Company enabled the 1878–1879 expansion from the original two-story 25- by 100-foot laboratory into a research and development complex with a large brick machine shop and a two-story brick office and library.

The Menlo Park laboratory became a model for other electrical inventors (notably **Alexander Graham Bell** and **Edward Weston**) and influenced the development of research facilities in the industry. It also provided the template for the even bigger laboratory that

Edison opened in **West Orange, New Jersey**, at the end of 1887. This complex included a three-story main building with experimental rooms, two machine shops, and a large library. It also had four small (25- by 100-foot) buildings, each with its own purpose (electricity, chemistry, and so on). At its height, Edison's West Orange staff was nearly three times that of Menlo Park. The surrounding area grew into a corporate complex that manufactured and marketed Edison's inventions.

Edison used several small labs between the closing of Menlo Park and the opening of West Orange. In September 1882, he moved

Sketch of Edison's Menlo Park laboratory complex by Richard Outcault (1880). The original wood-frame laboratory building is in the center. The brick office and library building is in the foreground; at the back is the brick machine shop. The experimental electric railway is at right. *Courtesy of Thomas Edison National Historical Park.*

Edison's laboratory complex at West Orange, ca. 1900. The main building at right contained his office, library, machine shops, and experimental rooms; at left are four smaller specialized laboratories. *Courtesy of Thomas Edison National Historical Park.*

his research to the top floor of **Bergmann & Company** in **New York City**. In June 1886, facing increasing competition in the lamp industry, he moved his laboratory to the factory of the **Edison Lamp Company** near Newark, where he focused on improving both his lamp and its manufacturing processes. Finally, Edison created a small laboratory at his winter home in **Fort Myers, Florida**, so he could work on experimental projects while there.

LATHROP, GEORGE PARSONS (1851–1898). A respected American writer, editor, and journalist, Lathrop had a nearly decade-long connection with Edison from 1885 as an interviewer, prospective business partner, and would-be coauthor. Lathrop gave Edison and his **phonograph** favorable press treatment, but his hopes of securing a place in the "amusement phonograph" business—a stillborn precursor to the **Automatic Phonograph Exhibition Company**—were in vain. His subsequent proposals for an Edison biography and a book (to be coauthored with **Samuel Clemens**) went nowhere. In mid-1890, having recently published in *Harper's Magazine* a sympathetic piece called "Talks with Edison," Lathrop suggested collaborating with Edison on *Progress*, a futuristic novel akin to Edward Bellamy's popular *Looking Backward, 2000–1887*. Edison agreed to provide imaginative technical ideas around which Lathrop would construct a story, and for a time he did so. However, within a year, Edison lost interest in the project and did not make time to discuss with his coauthor either his dozens of pages of notes or Lathrop's draft chapters. (Edison's notes, with Lathrop's comments and queries, are available on the Edison Papers Digital Edition.) Lathrop tried to keep his word to publisher Samuel McClure but by 1896 had to settle for serializing what material he had in American and British newspapers under the title "In the Deep of Time." Despite the frustrations on all sides over *Progress*, in 1891, Edison allowed Lathrop to publish an exclusive article revealing his **kinetoscope** to the public.

LATIMER, LEWIS HOWARD (1848–1928). Inventor and engineer, Latimer was born in Massachusetts to parents whose escape from slavery and subsequent legal peril became a cause for prominent abolitionists. Latimer taught himself drafting and then worked for a series of inventors and enterprises, including **Hiram Maxim** and the United States Electric Lighting Company. While employed by the latter, Latimer patented improvements in lamp manufacture, installed lighting plants, and helped set up a lamp factory in England. The **Edison Electric Light Company** hired him by 1885 based on his extensive knowledge of the electric lighting industry, especially the lamps of its competitors. Latimer worked as a draftsman in the engineering department and assisted with patent infringement suits against rival companies. From 1889 to 1911, he was in the legal departments of the **Edison General Electric Company** and its successor, **General Electric** (GE). In that period, he published *Incandescent Electric Lighting: A Practical Description of the Edison System (1890)* and became chief draftsman (from 1894) for the Board of Patent Control, a patent pool created by GE and the **Westinghouse Electric Company**.

Latimer was a founding member of **Edison Pioneers** and its only African-American member. He also wrote poetry, prose, and plays. Among his poems was one written in 1888 to mark Independence Day (the anniversary of his father's birth into slavery), which Edison promised to record and send to London with an early batch of phonograph cylinders. (The recording, if made, is missing.) The next year, he sent another poem to Edison, who reportedly "read it over very carefully and said it was d—m good" (*TAEB* 9, 225 n. 2).

LEFFERTS, MARSHALL (1821–1876). Lefferts, a native of **New York City**, became a telegraph engineer and entrepreneur after the Civil War, in which he attained the rank of colonel. As president of the **Gold and Stock Telegraph Company** (since 1870), he took Edison under his wing and encouraged the latter's work on **stock tickers**. Thus began a business relationship and friendship that lasted until Lefferts died, with the older man acting as a mentor to the aspiring inventor. Lefferts may have prompted Edison's first patent application in **automatic telegraphy**, and he backed his protégé's work on the **electric pen**. Lefferts also provided valuable advice on the **patent system** and introduced Edison to **Lemuel Serrell**, who became Edison's attorney. Edison assigned several patents to Lefferts and manufactured quotation instruments for the cotton trade, perforators, and other devices for Gold and Stock. Lefferts occasionally prodded his protégé "with a sharp stick," according to **Daniel Craig**, but Edison would write that in general Lefferts "treated me well," and the two grew close (*TAEB* 1:247, 643). When **William Orton** died in 1878, Edison told a reporter that "if I get to love a man he dies right away. Lefferts went first, and now Orton's gone, too" (*TAEB* 4:238 n. 1).

LIPPINCOTT, JESSE HARRISON (1843–1894).

An industrialist from the Pittsburgh, Pennsylvania, area, Lippincott bought Edison's North American rights to the **phonograph** in June 1888, shortly after acquiring similar rights to the rival **graphophone**. Having amassed wealth through a series of enterprises in groceries, baking, and glassmaking, he was based in **New York City** at the time but was unknown to Edison before **Ezra Gilliland** and attorney **John Tomlinson** brokered the deal. Lippincott created the **North American Phonograph Company** to market both the phonograph and the graphophone.

Edison was to receive $500,000 in cash installments from Lippincott (and retain his manufacturing rights). Under a side agreement, Lippincott was to pay half as much to Gilliland. Edison was furious when he learned this, but he and Lippincott strove to honor their contract. Lippincott and North American, hampered by design and production problems and a weak market, could not meet their obligations to Edison and the **Edison Phonograph Works**. Lippincott suffered a paralyzing illness in October 1890 and never recovered either his physical or his financial well-being.

LLEWELLYN PARK.

The longtime Edison home **Glenmont** is located in this planned community (one of the first in the United States) within the township of **West Orange, New Jersey**, about a dozen miles from **New York City**. Llewellyn Haskell planned the semirural village on his 425-acre estate in the 1850s, designing it in accord with the Romantic Landscape movement. **Mina Miller Edison** recalled years later that Edison gave her the choice of living there or on Riverside Drive in New York City. She chose the former because she thought her husband preferred the peaceful location as more conducive to his work. After Edison built his new **laboratory** in 1887, he could easily walk to work. He also lighted some of the neighbors' homes with power from the laboratory. He and Mina called Llewellyn Park home from 1886 to the end of their lives. Two of their children (**Charles Edison** and **Madeleine Edison**) made their homes there as well.

LONDON CRYSTAL PALACE EXHIBITION (1882).

"The *raison d'etre* of the Crystal Palace Exhibition," a prominent British engineering journal stated in its 17 February 1882 issue, "is to show that the electric light is not a failure, and that it daily promises to be as economical as the gas at present supplied" (*The Engineer* 53:126). The February–June 1882 exhibition in this vast London glass conservatory showcased the latest advancements in the practical use of electricity. **Edward Johnson** built the Edison exhibit (which opened in January) around materials brought over from the **Paris International Electrical Exposition (1881)**, including telegraphs, the **electric pen**, and the **phonograph**. But "above and surrounding all," Johnson wrote to Edison, "is your last contribution namely the **Electric Light**," which Johnson presented with characteristic showmanship (*TAEB* 6:381). Edison's lamps awed the Duke and Duchess of Edinburgh and other prominent visitors; the Edison light was, according to the *London Daily News*, "the wonder of the show" (*TAEB* 6:312). On the other hand, Johnson crowed, exhibits of rivals such as **Hiram Maxim** were "failure[s]," "nothing new," or "abortive" (*TAEB* 6:404). Johnson's work at the Crystal Palace (aided by **William Hammer**) overlapped with his planning and construction of the demonstration Edison central station at **Holborn Viaduct**, and both projects helped spur the creation of the Edison Electric Light Company Ltd of London in March 1882.

LOWREY, GROSVENOR PORTER (1831–1893). A well-placed New York attorney, Lowrey performed a range of official and unofficial services for Edison and his fledgling electric light companies. He met Edison in 1875 over patent matters for **Western Union Telegraph Company**, for which he was general counsel. Drawing on his connections with Western Union and with **Drexel, Morgan & Company**, Lowrey facilitated formation of the **Edison Electric Light Company** to finance the research and development of Edison's incandescent lighting system. He helped secure **New York City**'s approval for the **Edison Electric Illuminating Company of New York** to begin laying underground conductors for the **Pearl Steet Station ("First District")** in Manhattan in 1880–1881.

In 1882, Lowrey helped organize the Paris-based **Compagnie Continentale Edison** to license Edison's light and power patents in continental Europe. The next year, he aided negotiations with Drexel, Morgan, leading to the incorporation of Edison's manufacturing shops (including the **Edison Machine Works** and the **Edison Lamp Company**) as separate entities. From 1889 to 1891, Lowrey helped defend Edison's basic lamp incandescent lamp patents in two separate suits involving **Westinghouse Electric Company** (the so-called McKeesport and Filament cases). The **Edison General Electric Company** won each case and received judicial imprimaturs on broad interpretations of the Edison patents.

Lowrey was a native of Massachusetts and admitted to the bar in Pennsylvania in 1854. After involvement with the Free-State cause in Kansas before the Civil War, he became an Albany correspondent for the *New-York Evening Post*. He came to New York to practice law and was for many years a partner in Lowrey, Soren & Stone and its successors.

LUMIÈRE BROTHERS. Auguste (1862–1954) and Louis Jean (1864–1948) Lumière of Lyon, France, were manufacturers of photographic equipment who invented a combination movie camera/projector: the Cinématographe. The Lumières debuted this innovation at their Lyon factory in March 1895, anticipating the less efficient Latham projector (an American device that **William K. L. Dickson** helped develop) by several weeks. They presented the machine to the public in December 1895 by projecting a series of short films, including the now famous *L'arrivée d'un train à la Ciotat*, at the Grand Café in Paris.

The Cinématographe differed in several key respects from the kinetoscope. For one, its projection capability allowed multiple viewers to see film simultaneously, whereas the kinetoscope required individual viewers to peer, one at a time, through an eyepiece into a cabinet. The Cinématographe could photograph or project sixteen frames per second rather than Edison's forty frames per second. Another difference was portability. Powered by a hand crank instead of a heavy battery, the Lumières' device was light enough to be carried in one hand, greatly increasing the range of what might be filmed.

M

MACKENZIE, JAMES URQUHART (1837–1894). As a fifteen-year-old boy selling newspapers on the Grand Trunk Railway in the summer of 1862, Edison rescued the young son of MacKenzie, the station operator at Mount Clemens, Michigan, from an oncoming freight car. The indebted MacKenzie, a native of Scotland, offered his teenaged benefactor formal telegraphy lessons. During his apprenticeship, Edison developed proficiency with the telegraph, a skill that led to a stint as an itinerant telegrapher. As Edison's career flourished, he occasionally boosted MacKenzie's. For example, in 1887, Edison helped MacKenzie secure a contract with the city of Orange, New Jersey, for a fire alarm system. Edison also employed MacKenzie as an agent for the **electric pen** and on at least one occasion gave money to his hard-up friend. In turn, MacKenzie continued his role of mentor, advising Edison to "[be] temperate in all things . . . quit for meals regularly on the stroke of the clock . . . devote your evenings to domestic pleasure and enjoyment . . . [and] never work on Sunday!" (*TAEB* 3:644). He also occasionally visited (and tinkered at) Edison's **laboratory** in **Menlo Park, New Jersey,** and the two collaborated on a stencil **typewriter,** for which Edison received a patent in 1884. MacKenzie received several patents of his own between 1883 and 1890. After MacKenzie's death, Edison recalled him as "a kindly, honest man and a good friend" (TAE to William Hammer, ca. 7 December 1894; *TAED* X098A067).

MALLORY, WALTER SEELEY (1861–1944). Mallory had more than thirty years of business connections with Edison as a top executive and administrator in numerous enterprises. A Connecticut native educated in Baltimore, he was in the iron business in Chicago when he met Edison through a brother of **Mina Miller Edison** around 1885. (He ushered when Mina married Edison in 1886.) Mallory and Edison formed the Edison Iron Concentrating Company in 1888, with Mallory as secretary and treasurer. The Chicago-based company built and operated a trial ore-separating plant in Humboldt, Michigan, for about two years. After the plant burned in 1890, Mallory came east to help with other Edison ore ventures. He invested in and became vice president of the **New Jersey and Pennsylvania Concentrating Works** and from 1894 oversaw the company's plant at Ogdensburg, New Jersey. Edison entrusted Mallory with a succession of other important positions, including the founding president of the **National Phonograph Company** (1896–1899) and vice president (1899–1910?) and then president (ca. 1910) of the **Edison Portland Cement Company.** Mallory was also an official or director in other Edison companies, including the Edison Crushing Roll Company, the Mining Exploration Company of New Jersey, and the Edison Electric Light Company of Europe. He retired in 1918 but continued doing statistical research for the Portland Cement Association until 1943. He served many years as treasurer of the **Edison Pioneers.**

MAMBERT, STEPHEN BABCOCK (1887–1958). Mambert was a young efficiency expert who, with **Charles Edison**, reorganized **Thomas A. Edison Incorporated** (TAE Inc.) in 1915. Mambert had studied management engineering at Cornell and was hired as a clerk at TAE Inc. in 1913. When Edison was rebuilding his **West Orange, New Jersey**, factories after the calamitous 1914 fire, he wished to bring in new production processes pioneered by **Henry Ford**. Mambert seized this opportunity, providing a plan to rationalize TAE Inc.'s operations from top to bottom. With Charles Edison, he reshaped the sprawling organization into a set of semi-independent divisions linked by a central bureaucracy for common functions, such as procurement. He standardized procedures, took away traditional prerogatives of factory foremen, and challenged the personal fiefdom of **Miller Reese Hutchison**. Mambert's rise at TAE Inc. was meteoric. He became the in-house "efficiency engineer" in March 1915, reporting only to Edison and the general manager, and two months later was a vice president. His fall was also rapid. He lost Edison's confidence (for uncertain reasons) and resigned in 1924. He did not return to the Edison orbit.

MARTIN, THOMAS COMMERFORD (1856–1924). Although Martin was an important associate and friend of Edison, it remains unclear when the two first worked together. Obituaries claimed that he entered Edison's service in 1877, but his name first appears as an employee of the **Edison Speaking Phonograph Company** in 1878. He subsequently made his name as editor of the *Electrical World*. Martin authored numerous publications, including *Inventions, Researches, and Writings of Nikola Tesla* (1893) and (with **Frank Dyer**) the authorized biography *Edison: His Life and Inventions* (1908). One of the founding members of the **American Institute of Electrical Engineers**, Martin served as its first acting secretary and later as president. He was secretary of the National Electric Light Association from 1909 to 1921 and an adviser thereafter.

MATTER (EDISON'S CONCEPTIONS OF). Edison did not articulate a coherent view of the fundamental structure of the material world (which mattered less to his day-to-day work than did the nature of energy), but he alluded to underlying beliefs and speculated beyond his practical need for knowledge. Edison's supposition that matter is composed of atoms and molecules was a conventional one. However, his deistic belief in an infinite and timeless universe "because the mind of man cannot conceive the creation of so much matter," which he probably absorbed from **Thomas Paine**, put him at odds with Christian teachings (*TAEB* 4:267). So too did the correspondence that he (and contemporary scientists) postulated between the worlds of the very large and the very small: "Our solar system is a Cosmical Molecule," he suggested in 1886 (*TAEB* 8:487). Edison sometimes ascribed to matter properties that could help explain phenomena that could be described but not understood causally. These expressions further illustrate his indebtedness to **Michael Faraday**'s worldview of forces and charges. In 1878, just at the start of his electric light research, he proposed that "gravitation is not really a force it is only a condition of matter," a belief he also held of magnetism (*TAEB* 4:845). Later, with more experience, he proposed that the physical resistance to a wire moving through a magnetic field was due to "matterless friction" in the field—the "friction of Molecules"—condensing the invisible ether (*TAEB* 9:72). A more convincing explanation awaited knowledge of subatomic particles and charges, but Edison at one time assumed all atoms to have north–south polarities. These opposing charges, he thought, could explain gravity as the "mutual electrical attraction of all the atoms" (*TAEB* 8:487).

Perhaps the closest Edison came to a consistent view of matter was his enduring belief in an intelligence inherent in the smallest particles. This belief was consistent with panpsychism—a view of nature as endowed with the properties of mind—which has a long history in Western philosophy. As he put it in an 1878 aphorism, "Each ultimate particle of matter is endowed with intelligence" (*TAEB* 4:267). Edison elaborated in an 1885 interview with a friendly journalist:

I do not believe that matter is inert, acted upon by an outside force. To me it seems that every atom is possessed of a certain amount of primitive intelligence. Look at the thousand ways in which atoms of hydrogen combine with those of other elements, forming the most diverse substances. Do you mean to say that they do this without intelligence? When they get together in certain forms they make animals of the lower orders. Finally they combine in man, who represents the total intelligence of all the atoms. (*TAEB* 8:132)

Edison allowed that this intelligence originated with a Creator or God whose existence "can almost be proved from chemistry" (*TAEB* 8:132). However, in a 1910 interview, his belief in this primordial intelligence led him in a different direction. He appeared to question the existence of the Christian God, and he even raised the possibility of communicating with intelligent particles after the death of the human body. These suggestions caused a storm of protest. *See also* CHEMISTRY; ELECTROMAGNETISM; RELIGION AND RELIGIOUS VIEWS.

MAXIM, SIR HIRAM STEVENS (1840–1916). Maxim was an American inventor and engineer who had a short but bitter rivalry with Edison. He began making his own electric generators and arc lamps in New York, and the United States Electric Lighting Company was formed in 1878 to sell his products. Maxim also created incandescent lamps using a thin platinum strip (1877) and a thick carbon rod (1878), but neither was a practical device. In late 1880, he debuted a lamp with a flat carbonized filament (bent into a distinctive "M" shape) in a vacuum. Edison was incensed by the obvious resemblance to his own lamp and pointed out that Maxim had toured his **laboratory** in **Menlo Park, New Jersey,** and hired away glassblower **Ludwig Böhm.** The resulting dispute strained Edison's relations with physicists **Henry Morton, George Barker,** and **Henry Rowland** and played out at the **Paris International Electrical Exposition (1881),** where

both he and Maxim won honors. However, the commercial appeal of Maxim's lamp was overmatched by Edison's integrated **electric lighting system** as well as by the financial and legal advantages provided by the **Edison Electric Light Company.** Still, Maxim's patented process for "flashing" filaments with a uniform layer of carbon proved of lasting value. Maxim abandoned his American career (and his family) in the early 1880s and started over in Great Britain, where he improved the machine gun and invented (with his brother) smokeless gunpowder. He was knighted in 1901.

MAXWELL, JAMES CLERK (1831–1879). A Scottish mathematician and physicist, Maxwell was among Britain's foremost scientists of the nineteenth century. His influence only grew after his early death, as physicists and engineers digested his abstruse *Treatise on Electricity and Magnetism.* Their interpretation of Maxwell's work showed the explanatory power of **Michael Faraday**'s concept of the electromagnetic field regarding the transmission of energy (such as electricity and magnetic force) through space. The combined efforts of Maxwell and his acolytes put **electromagnetism** at the center of physics research in the English world in the 1880s, and efforts to understand it percolated through the scientific and technical publications that Edison read. Predisposed to conjecture about energy and its transmission through space, Edison seems to have drawn encouragement to speculate from the intellectual environment around Maxwell's work. In the mid-1880s, he hypothesized at length about energy and its convertibility from one form to another in a manner consistent with field theory. Edison also developed what might be called a research program of thought experiments and designed related physical experiments to try, though he did not carry them out in any meaningful way.

The Maxwellian enterprise in physics suggested a conceptual framework for Edison to understand how electrical signals appeared to jump through space, a phenomenon that he and **Ezra Gilliland** harnessed in their **railway ("grasshopper") telegraph** system. And it almost certainly nurtured his long interest in the existence of **unknown natural forces**—forms of energy not yet detected or named.

In addition to his studies in electromagnetism, Maxwell did pioneering theoretical work on the diffusion of gases. That topic had practical applications to telegraphy, where battery life was limited in part by the migration of hydrogen gas through the electrolytic fluid to the negative electrode, where it impeded the flow of current. When Edison made a significant discovery that addressed this problem, he considered sending a written account to Maxwell. However, he did not, and there is no known correspondence between the two men.

MCCOY, JOSEPH F. (1860–1938). McCoy conducted surreptitious investigations of personal and business matters for Edison. Having worked from 1880 to 1883 for the **Edison Lamp Company** and its predecessor, he returned to Edison's ambit in 1892 under the New Jersey Phonograph Company. For nearly fifty years afterward, McCoy's duties included service as an undercover agent and industrial spy for the Edison interests. He watched competitors in the **phonograph, motion picture,** and other businesses and sometimes surveilled Edison employees. McCoy also checked up on Edison's wayward sons **Thomas A. Edison Jr.** and **William Leslie Edison.** He reported on Tom's dubious attempts to monetize his father's name, and he rented a house near William's residence to keep a close eye on his activities. McCoy's loyalty to Edison was rewarded by his election to the **Edison Pioneers.**

MEADOWCROFT, WILLIAM HENRY (1853–1937). Born in Manchester, England, Meadowcroft came to the United States in 1875. He served as a law clerk under **Sherburne Eaton,** whom he followed to the **Edison Electric Light Company** in 1881 as an assistant. Thereafter, he held executive and administrative posts in several Edison enterprises, including the **Edison Company for Isolated Lighting,** the **Edison Ore Milling Company,** and the Edison Electric Light Company of Europe. From 1908 to 1910, Meadowcroft helped **Thomas Martin** and **Frank Dyer** prepare their official biography. In 1910, he became Edison's assistant and confidential secretary, a position he held until the inventor's death. Meadowcroft's fifty-year employment with Edison earned him monikers such as "Edison's Prime Minister" and "The Wizard's Boswell," and he once commented that "[Edison] and I understand each other before we have spoken." Like other loyal Edison employees, he put in long hours: "I used to get acquainted with my family on Sunday," he once reminisced (obituary, *New York Times,* 16 October 1937, 19). Meadowcroft wrote several books of his own, including *The ABC of Electricity* (1888), *The ABC of the X-Ray* (1896), and *The Boys' Life of Edison* (1911). A founding member of the **Edison Pioneers,** he briefly served as the group's president. Meadowcroft was buried on 18 October 1937—the sixth anniversary of Edison's death.

MENLO PARK LABORATORY IN GREENFIELD VILLAGE. In October 1929, Edison's friend **Henry Ford** opened the Edison Institute in Dearborn, Michigan, as a culminating event of Light's Golden Jubilee, celebrating the fiftieth anniversary of Edison's incandescent **electric lamp.** The Institute (known today as The Henry Ford) included both a large museum—modeled on Independence Hall—and Greenfield Village, the centerpiece of which was a reconstruction of Edison's **laboratory** in **Menlo Park, New Jersey.** Ford had planned to move the original laboratory buildings from New Jersey, but all that was left of Edison's property were foundations and the glass house (formerly a glassblowing shop, drafting room, and photographic studio preserved by the **General Electric Company** and subsequently donated to the Institute). The **Sarah Jordan boardinghouse** also survived, and Ford moved it to Greenfield Village and put it across the street from Edison's laboratory. Next to it, he placed Edison's original laboratory from **Fort Myers, Florida.** Ford furnished the buildings with period machines and equipment, including many items donated by Edison, such as patent models from the **Newark, New Jersey,** and Menlo Park periods. He also acquired a significant collection of other Edison-related artifacts and original Edison documents, now housed in the Benson Ford Research Center.

MENLO PARK, NEW JERSEY. In December 1875, Edison purchased two tracts of land and a house in the hamlet of Menlo Park, a failed real estate development twelve miles south of **Newark, New Jersey**, and a whistle-stop on the Philadelphia–**New York City** railroad line. There he erected a new **laboratory** where he would produce his **phonograph** and **electric light system**. Edison was attracted to Menlo Park by its setting, which he described in a letter to a friend as "the prettiest spot in New Jersey" (Israel 1998, 123). It provided his young family a detached home and yard on a sizable tract of land some distance from Newark and New York City, which, like other growing American cities, were seen as crowded, unhealthful, and dangerous.

Edison and his family lived in Menlo Park for five years before moving in March 1881 to New York City, where he could better supervise installation of his **Pearl Street Station ("First District")** in Lower Manhattan. The family continued to live in their Menlo Park home during the hot summer months. Edison kept the laboratory open until September 1882, then used it for occasional experiments and for planning central stations. He largely abandoned the property after the unexpected death of his wife **Mary Stilwell Edison** in August 1884, though members of her extended family lived there intermittently as caretakers until 1891. The laboratory buildings were put to various other uses until the 1910s, when local residents dismantled them for building materials. Fire destroyed the house in 1917. The site of Edison's laboratory is now part of Edison State Park. *See also* THOMAS EDISON CENTER AT MENLO PARK.

MICROPHONE CONTROVERSY. Edison cultivated the respect of influential engineers and scientists in Great Britain, but his ire in 1878 about what he saw as an egregious appropriation of inventive credit caused him to breach the protocols of those fraternities. The transatlantic dispute began in May, when Edison read accounts in British journals about inventor David Hughes's research on amplifying sound with a device he called the microphone. The articles credited Hughes with discovering that the resistance of semiconducting substances, such as carbon, varies with pressure in a way that could be used to amplify a signal. They also suggested that the Hughes microphone was an improved form of the **carbon-button telephone transmitter** that Edison claimed as his own. He became especially upset after reading Hughes's paper to the Royal Society acknowledging help from **William Preece**. Edison could not understand why Preece, who had visited Edison's **laboratory** in 1877 and subsequently helped introduce his telephone into Britain, had not pointed out to Hughes that this was "nothing more than my carbon telephone" (*TAEB* 4:287). And he was enraged when another article credited Hughes with adapting the discovery to a thermopile for measuring heat, which Edison had already done with his **tasimeter**. He cabled Preece, "I regard Hughes heat measurer & Direct Impact telephone as abuse confidence. I sent you & others papers describing it . . . if you do not set this thing right I shall with details" (*TAEB* 4:327). He also cabled his allegations about Preece's conduct to eminent physicist **Sir William Thomson**.

The dispute became public when Edison responded to an editorial in the *New York Daily Tribune* arguing that Hughes's microphone differed from the carbon telephone. In his letter to the editor and in subsequent newspaper interviews, Edison elaborated his claims and the accusations against Preece. Assisted by **Charles Batchelor**, he sent copies of his *Tribune* letter and supporting materials to leading figures in the British scientific community and some European luminaries. The dispute became a matter of national pride, as newspapers and scientific journals in the United States and Britain framed the question of scientific credit in those terms. Because he made the matter a personal crusade against Preece and pressed his claims through the popular press, Edison ultimately failed to convince the British scientific community to see his side or concede him a share of the credit. The affair probably informed the skeptical reaction British scientists gave to Edison's claims for his electric light system a few years later, particularly regarding his English rival **Joseph Swan**. The rift with Preece did heal after several years.

MILAN, OHIO. Founded in 1817, this town in northwestern Ohio began to prosper with the 1839 completion of a three-mile canal connecting it to the Huron River. The economic opportunities expected to follow from the canal drew **Nancy Edison** and **Samuel Edison** to Milan in 1838–1839, and by the time Thomas was born there in 1847, the village had become one of the busiest harbors on the Great Lakes and exported more wheat than any port in the world except Odessa, Russia. Milan was also a regional center of shipbuilding and stave manufacture. Yet the exhaustion of area timber and the construction of railroads nearby abruptly brought economic obsolescence, and in 1854, the Edison family moved to **Port Huron, Michigan**. Edison remembered little of his time in Milan and made few return trips. His last visit came in August 1923, when a party including **Henry Ford** and **Harvey Firestone** dropped in after the nearby funeral of Warren Harding. Edison was chagrined to find that his boyhood home, then occupied by a distant cousin, lacked electricity. Today, the home, a National Historic Landmark listed on the National Register of Historic Places, is managed as the **Thomas Edison Birthplace Museum**.

MILLER, JOHN VINCENT (1873–1940). A younger brother of Edison's second wife, **Mina Miller Edison**, Miller was closely connected with Edison and his businesses. After attending Yale University, he enlisted in the U.S. Navy during the Spanish-American War. He began working for Edison in 1899, overseeing construction of an experimental gold ore processing mill (using Edison's dry placer process) at the Ortiz Mine in Dolores, New Mexico. The following year, Edison sent him to the Sudbury district of Ontario, Canada, to prospect for nickel and cobalt for use in storage batteries. Miller then served for several years as agent of the Mining Exploration Company of New Jersey, the company Edison organized in 1902 to finance his ongoing search for these ores. He subsequently held managerial or executive positions in several Edison companies (including the Edison Chemical Works), served as Edison's personal business secretary in the 1920s, and acted as executor of Edison's estate.

MILLER, LEWIS (1829–1899). The father of Edison's second wife, **Mina Miller Edison**, Miller was an Ohio industrialist and philanthropist. He was a partner in Aultman, Miller and Company of Akron, makers of agricultural machinery, such as the Buckeye mower, and he received several patents on cutting and harvesting equipment. Despite their different views of **religion**, Miller was sympathetic to Edison, perhaps informed by his own experiences as a manufacturer and inventor. He also tried to assuage Mina's doubts about her role as a young wife and stepmother. A devout Methodist, Miller had a deep interest in education that led to him to create the so-called Akron plan of church architecture, which dedicated space for religious instruction. He cofounded the **Chautauqua Institution** in 1874 to provide a mix of education, leisure, and moral uplift and served as its president until his death. He served as president of the Methodist-affiliated Mount Union College in Ohio. Among the eleven children born to Miller and his wife **Mary Valinda Miller** was **John Vincent Miller**, who had his own business relations with Edison.

MILLER, MARY VALINDA (née Alexander, 1830–1912). Miller was the mother of Edison's second wife, **Mina Miller Edison**. A native of Illinois, she married Ohioan **Lewis Miller** in 1852. Mary Valinda's voluminous correspondence with her children provides much information about activities and events in the extended family, including Edison and Mina. She and Lewis lived for decades at Oak Place, a bucolic estate overlooking the industrial city of Akron, Ohio. **John Vincent Miller** was also among their eleven children.

MILLER, WALTER HENRY (1870–1941). Miller was a local boy from East Orange, New Jersey, who joined the staff of the **laboratory** in **West Orange, New Jersey**, in 1887 as a machine shop apprentice and spent his entire career with Edison and his enterprises, mostly in sound recording. Miller quickly became one of Edison's **phonograph** experts. He helped develop an improved recording stylus for the cylinder phonograph, a reproducer that did not require adjustment, and a thermal process for making a long succession of duplicate wax cylinders inside durable metal master molds. Miller and **Jonas Aylsworth** took out the essential patents on that process and oversaw its implementation in a duplicating plant that, a little more than a year after its inauguration, was turning out 10,000 "Gold-Molded" records a day in 1902 to meet demand. A few years later, Edison gave Miller and Aylsworth the task of manufacturing a cylinder that would play four minutes—as long as the discs of the rival **Victor Talking Machine Company**— and gave each man $10,000 on the project's completion. When Edison belatedly decided, at the end of 1909, to produce his own disc phonograph and recordings—the Edison Diamond Disc—he again turned to Miller (then head of the **National Phonograph Company**'s recording department) to create a manufacturing process for the records. After the formation of **Thomas A. Edison Incorporated** in 1911, Miller headed its Recording Division until it ceased producing records in 1929. He then transferred to the Silver Lake, New Jersey, factory complex to oversee production of cylinders for the Ediphone dictating machine. Miller retired in 1937.

MIMEOGRAPH. The mimeograph, also known as the stencil duplicator, was a commercially successful system for copying documents devised by lumber merchant **Albert Dick** in the mid-1880s. Dick created stencils by tracing the original with an electric perforating needle, then squeezing ink through the tiny holes onto a sheet of paper below to make a copy. Unknown to Dick, Edison had already patented a similar perforating device for his **electric pen**. Edison also had a separate patent on a method of creating typed stencils. The two men reached a licensing agreement in 1887 that allowed Dick to make and sell the device as Edison's Mimeograph. Edison improved the ink, paraffin stencil paper, and typewriters for cutting stencils, but he was not directly involved in producing the machines. The Edison–

Dick Mimeograph (as it was known from about the time of World War I) remained standard equipment in U.S. offices and schools into the latter half of the twentieth century.

MORGAN, JOHN PIERPONT (1837–1913). American financier J. P. Morgan was the controlling force behind the **Drexel, Morgan & Company** investment bank and himself a dominant figure in the U.S. economy in the late nineteenth and early twentieth centuries. Morgan rose to his preeminent position by financing the railroad boom in the 1870s and 1880s and later increased his power by investing in the steel, machinery, and electrical industries and by refinancing U.S. government debt. His decisive actions in the Panic of 1907, when he used his personal and business wealth to stop a run on banks and brokerages, prompted congressional inquiries and contributed to the creation of the Federal Reserve in 1913.

Morgan was enthusiastic about emerging technologies, including electric lighting, particularly after he visited Edison's **laboratory** in **Menlo Park, New Jersey**, in December 1878. His firm quickly acquired most European rights to Edison's lighting patents and began financing the establishment of Edison companies in London and Paris. At considerable risk, Morgan steadily supported Edison's lighting research, even when it foundered. Although his early financial commitments were through his bank, he eventually made large personal investments in Edison lighting enterprises. He also gave his personal imprimatur to the new system by having a lighting plant installed at his Manhattan mansion in 1881, the first private residence in the world so outfitted. When Edison started the **Pearl Street Station** (**"First District"**) plant in Lower Manhattan the next year, Morgan's office was among the first set aglow.

Despite this support, Morgan acted cautiously and in general favored consolidation and limited competition. Edison, by contrast, wanted to expand quickly and crush potential rivals. These differences came to a head during negotiations over consolidating Edison's manufacturing shops and the **Edison Electric Light Company** into the **Edison General**

Electric Company (EGE) in 1888–1889. Although Morgan stayed in the background of EGE, he seems to have had a major part in pushing it into the 1892 merger that created the **General Electric Company**.

MORTON, HENRY (1836–1902). Morton, a professor of chemistry, was president of the Stevens Institute of Technology in Hoboken, New Jersey, from 1870 until his death. He was known for his theatrical public lectures and service as a paid consultant in patent disputes. Morton arranged for Edison to conduct etheric force experiments at Stevens in 1875, and the two men occasionally met and corresponded afterward. Morton witnessed a **phonograph** demonstration in **New York City** and joined Edison on a solar eclipse expedition in 1878. He was among the scientists from whom Edison sought to borrow a vacuum pump for his nascent electric light experiments, but their relationship soured over the course of that research. Morton publicly questioned the utility of the prospective Edison **electric light and power system**. In 1879, he called it "a conspicuous failure trumpeted as a wonderful success," hinting that Edison had duped a newspaper reporter (Bazerman 1999, 186). Over time, Morton endorsed the rival electric system of **Hiram Maxim**, coauthored a *Scientific American* article disparaging the efficiency and practicality of Edison's system, and criticized the safety record of Edison's underground wiring. Edison scorned Morton as "ignorant" of electric lighting and lamented in a letter to him that "you have not treated me exactly right for reasons [which] I cannot fathom" (*TAEB* 5:906). Long after Morton's death, Edison referred to him as "my worst enemy," who "did every thing possible he could for the gas [companies] to make it hard for me to introduce the Electric Light" (TAE marginalia on letter from Henry Morton [son], 31 December 1918, *TAED* E1817BC).

MOTION PICTURE CAMERA (KINETO-GRAPH). In the 1880s, several inventors were working on cameras to produce **motion pictures**, including Edison and his assistant **William K. L. Dickson**. Together, Edison and Dickson developed the kinetograph camera in 1891, with Edison taking out the patent (U.S. Patent 589,168). On the basis of that patent, Edison was awarded priority in later lawsuits and received wide credit for the invention. His device was powered by an electric motor and used an escape mechanism to advance sprocketed film in front of the lens at about forty frames per second in stop-and-start fashion. These short films were viewed with the Edison **kinetoscope**. Edison named the kinetograph—among other inventions—using an original combination of Greek roots.

MOTION PICTURE PATENTS COMPANY (MPPC). This company was organized in 1908 by the **Edison Manufacturing Company** and the **American Mutoscope and Biograph Company** to acquire, pool, and license patents relating to the manufacture of motion pictures, which had been the subject of lawsuits between those parties. MPPC subsequently entered into price, royalty, licensing, and other agreements with other producers, importers, rental exchanges, exhibitors, and manufacturers as well as with the Eastman Kodak Company. It licensed exclusive distribution rights to the General Film Company, which was incorporated in April 1910 and controlled by MPPC. MPPC's distribution practices became an important model for Hollywood, but it was unable to establish firm control over the domestic market or prevent competition from independent producers and exchanges.

Frank L. Dyer, vice president of the Edison Manufacturing Company and later president of **Thomas A. Edison Incorporated**, was MPPC's founding president. Harry N. Marvin, president of American Mutoscope and Biograph, served as vice president and succeeded Dyer as president in 1912. George F. Scull, assistant to the vice president of Edison Manufacturing, was secretary. Other motion picture companies represented on the MPPC board of directors included Pathé Frères, Georges Méliès, Kleine Optical Company, Kalem Company, Lubin Manufacturing Company, Vitagraph Company of America, Selig Polyscope Company, and Essanay Film Manufacturing Company.

The federal government filed an antitrust suit in August 1912 that resulted in rulings against the company in October 1915 and January 1916. MPPC appealed to the U.S. Supreme Court, but after a settlement was reached, the appeal was withdrawn, and the company was dissolved.

MOTION PICTURES. Edison began working on motion pictures after attending a lecture by **Eadweard Muybridge**, whose zoopraxiscope used a series of photographs made in rapid succession to display the motion of animals. Stimulated by his discussion with Muybridge, Edison sought to preserve and reproduce motion by creating "an instrument that does for the eye what the phonograph does for the Ear" (*TAEB* 9:395). His first design involved photographing a series of small pictures, mounting them on a revolving cylinder like that of his **phonograph**, and viewing them as they turned through the enlarging eyepiece of a microscope. He planned to create sound movies by linking the device with a phonograph.

Edison relied heavily on **William K. L. Dickson**, a member of his experimental staff who was also a photographer. While Edison provided the resources, the vision for the invention, and the electromechanical knowledge, Dickson provided most of the photographic expertise needed for motion pictures. By 1892, the two had developed a camera called the kinetograph and a peephole viewing device known as the **kinetoscope**. Production of motion pictures began the next year in a custom studio (nicknamed the Black Maria) at the **laboratory** in **West Orange, New Jersey**. Its walls were covered in tar paper, and it rested on a circular track in order to follow the sun to admit light through its hinged sunroof. The first public viewing of motion pictures took place in May 1893 at the Brooklyn Institute of Arts and Sciences. The first kinetoscope parlor opened to the public the following year.

Edison was the first to introduce a commercial motion picture system, but he played almost no role in the development of projector technology and other commercial improvements. Indeed, although the first projector used by the Edison film company was called the Edison Vitascope, it was designed by C. Francis Jenkins and Thomas Armat in 1894. Dickson left Edison's employ in 1895 and helped to establish the rival **American Mutoscope and Biograph Company**. Edison placed the production of motion pictures under the direction of **William Gilmore**, general manager of the **Edison Manufacturing Company**. Edwin S. Porter, who joined the company in 1900, became a director and made some innovative films, including *The Life of an American Fireman* and *The Great Train Robbery*. Porter left in 1909 after **Frank Dyer**, Gilmore's successor, demoted him.

In the early 1910s, Edison and his laboratory staff made significant efforts to develop two motion picture technologies, but both failed commercially. The first was the home kinetoscope, a small projector for showing movies in homes, schools, and churches. Edison also produced educational movies for this device and advocated for their use in schools. The second was the **kinetophone**, a system to mechanically synchronize film and phonograph recordings.

Edison Manufacturing joined with American Mutoscope and other leading firms in 1908 to create the **Motion Picture Patents Company (MPPC)**—an unsuccessful effort to control film technology, production, and distribution. Independent competitors continued to thrive between 1908 and 1915, when MPPC lost a federal antitrust suit. Edison's own motion picture business became a division of **Thomas A. Edison Incorporated** in 1911 and was sold to the Lincoln and Parker Film Company in 1918. *See also* MOTION PICTURE CAMERA (KINETOGRAPH).

MURRAY, JOSEPH THOMAS (1834–1907). Murray was an early manufacturing partner of Edison. He met Edison while working as a machinist in **Newark, New Jersey**, first with the **American Telegraph Works** and then at Edison & Unger. By January 1872, Murray began witnessing entries in Edison's notebooks; in February, they established Murray & Company, a small Newark shop that made telegraph instruments. After Edison ended his association with **William Unger** that summer, he and

Murray created the manufacturing firm of Edison & Murray and occupied the Ward Street shop formerly used by Edison & Unger. They expanded Edison's **laboratory** in May 1875 to encompass the top room of the building along with half of another floor and a quarter of the cellar. Yet Edison and Murray dissolved their partnership in July 1875, and Edison moved to his new laboratory in **Menlo Park, New Jersey**, in early 1876. Murray continued to manufacture telegraph instruments at Ward Street and made the first Edison telephones there in 1877–1878.

MUYBRIDGE, EADWEARD (1830–1904).
Muybridge was an English-born photographer who did pioneering work in the photographic study of motion (chronophotography) in the United States. Born Edward James Muggeridge, he changed his name to what he believed was an authentic Anglo-Saxon form. He immigrated to California in 1852 and achieved wide fame in 1868 with his large-format photos of the Yosemite Valley.

Muybridge began his motion studies in 1872 with photographs of a galloping horse. He later expanded his work to other animals and humans. His invention of the zoopraxiscope (1879) contributed significantly to the early development of motion pictures. The instrument projected a series of drawings printed on a rotating glass disk to create the illusion of movement. The drawings were derived from still photographs taken by a battery of cameras set up to capture motion.

Muybridge came to Orange, New Jersey, in February 1888 to lecture on animal locomotion before the local New England Society. Edison very likely attended this talk. Muybridge visited Edison's laboratory two days later and subsequently said that he and Edison discussed combining the zoopraxiscope with the phonograph, though Edison denied having such a conversation. Muybridge returned to Orange in May 1888 to lecture and again conferred with Edison, who requested some plates from Muybridge's recent book *Animal Locomotion*. Edison received 100 plates and reportedly set them up around his library for study. **Walter Mallory** later attributed the start of Edison's **motion picture** experiments to these photographs.

N

NATIONAL ACADEMY OF SCIENCES
(NAS). In 1863, the U.S. Congress created the NAS as a private advisory board for military technology. Following the Civil War, the Academy functioned as an honorary society of sorts that provided recognition and support to promising scientists and inventors. Among the latter was Edison, who in November 1874 demonstrated his **electromotograph** before Academy members in Philadelphia. ("They all agreed that it was [an] original and important discovery," Edison reported to his father [*TAEB* 2:331].) Throughout the 1870s and 1880s, the NAS invited Edison to its gatherings, and his April 1878 demonstration of the phonograph before an Academy meeting at the Smithsonian brought him considerable publicity. He debuted a voltage-smoothing device at an Academy soiree in November 1882, and later that month, he invited members to visit his new **Pearl Street Station** (**"First District"**). Yet the Academy's enthusiasm for Edison and his work eventually waned, as the growing power and prestige of theory-based scientific research weakened the bonds between inventors and the science community. In 1911, Edison's nomination for membership received only three votes. Four years later, Edison deliberately omitted the group from a list of engineering societies to be solicited for the **Naval Consulting Board**, as he wished the Board's members to have a practical bent rather than be inclined to theory. The NAS finally elected Edison to membership in 1927. *See also* SCIENCE (EDISON AND).

NATIONAL PHONOGRAPH COMPANY.
This firm was incorporated in New Jersey in 1896 to succeed the **North American Phonograph Company** and restore to Edison control of the **phonograph** business. Edison had long been dissatisfied with his 1888 sale of phonograph rights in North America to investor **Jesse Lippincott** and the North American Phonograph Company, and he took control of North American after Lippincott died in 1891. He forced the company into bankruptcy in 1896 and organized National Phonograph to take over and expand its operations. **William Gilmore**, general manager of the **Edison Manufacturing Company**, became president of the new enterprise. Its profits were assigned to Edison in exchange for making technical improvements in the phonograph. National Phonograph also manufactured records for home entertainment, prompting Edison to make innovations in recording methods and duplicating technology, which by 1911 included disc recordings. Meanwhile, the company's foreign department (which also supervised the overseas interests of the Edison Manufacturing Company and the Bates Manufacturing Company) operated distribution offices, recording studios, and factories for making records in Great Britain, Germany, France, Belgium, Australia, and Mexico. National Phonograph was reorganized in February 1911 and became the core of the new **Thomas A. Edison Incorporated**.

NAVAL CONSULTING BOARD.
The sinking of the British passenger liner *Lusitania* by a German submarine on 7 May 1915 caused the

deaths of 1,198 people (including 128 Americans) and spurred a war preparedness movement in the United States. Within weeks of the tragedy, Edison proposed forming "a great research laboratory, jointly under military and naval and civilian control" (*New York Times*, 30 May 1915, SM6). His comments prompted **Miller Reese Hutchison**, his chief engineer, to suggest a board of civilians to advise the federal government on naval warfare technology. Hutchison, who viewed such a body as a means of getting Edison's **battery (storage)** into submarines, drafted a letter for Navy Secretary **Josephus Daniels** to send to Edison soliciting the inventor's help. Daniels's reworked version of the letter asked Edison to lend his "inventive genius" and advise the board (Daniels to TAE, 7 July 1915, *TAED* X021A). Edison agreed in July 1915 not only to serve but also to lead the twenty-four-member Naval Consulting Board, though he delegated the administrative responsibilities. His formal association with the Board lasted more than five years and proved deeply disappointing. The Board's primary task was to mobilize the nation's inventors and engineers (academic scientists being largely excluded, at Edison's suggestion), but of more than 100,000 inventions brought forward, only 100 received further investigation and only a single device was built. Edison's insistence on creating a naval research laboratory in Sandy Hook, New Jersey ("as far from Washington as possible" [Morris 2019, 193]), under civilian leadership was rebuffed, as were his suggestions for dozens of new military devices, including a self-stabilizing shell, a submarine searchlight, and a waterproof microphone. His tenure with the Consulting Board was also marred by a deadly explosion aboard the submarine *E-2* in January 1916 that a court of inquiry blamed on hydrogen gas from a faulty Edison battery. Edison quit the Board in January 1921 after the navy decided to build its new Naval Research Laboratory in the nation's capital. *See also* WORLD WAR I RESEARCH.

NEW JERSEY AND PENNSYLVANIA CONCENTRATING WORKS (NJPCW). Edison

incorporated the NJPCW in December 1888 to engage in iron mining, ore separation and concentration, smelting, and the sale of iron in New Jersey and Pennsylvania, where the iron industry was in long-term decline. The company could acquire mining properties and license Edison's patents from the **Edison Ore Milling Company** (and acquire those of other inventors). Edison served as president, with operational duties delegated to a general manager and a secretary; William Mallory was vice president. Edison and financier **Robert Cutting Jr.** owned most of its stock, though Edison associates William Perry (secretary of the **Edison Ore Milling Company**), **Samuel Insull**, and **Alfred Tate** held small stakes. In 1889, NJPCW built and briefly operated an experimental ore milling plant using Edison magnetic separators at Bechtelsville, Pennsylvania, to demonstrate the efficacy of the Edison machine to mine owners. The plant's initial promise led Edison to expand it and go into the mining and ore milling business for himself. The enlarged Bechtelsville plant, lacking enough suitable ore, fizzled after a few months, and Edison closed it in January 1890. He shifted the machinery to Ogdensburg, New Jersey, where he and the NJPCW had bought or leased a cluster of iron mines. The NJPCW built a sprawling complex there for mining and ore separation and concentration. Edison devoted most of his time to the Ogdensburg project from 1890 to 1901. He twice rebuilt and expanded the plant, which came to include (in addition to his magnetic separators) his patented giant crushing rolls, water separators, bricking machines, and other novel equipment. He eventually created a fully automated system in which rock, ore, and derivative products like sand moved on conveyor belts. The system worked but required some 400 hands to operate and keep it in repair. The resultant high labor costs, combined with changing market conditions for Bessemer ore, forced Edison to close the plant for good in December 1900—reportedly after spending $2.2 million of his own money there. Edison later applied some of the key processes and machinery to the manufacture of cement. *See also* EDISON PORTLAND CEMENT COMPANY; MALLORY, WALTER SEELEY; ORE (IRON) MILLING.

NEW YORK CITY. Edison spent most of his adult life in or near New York, which consisted primarily of Manhattan until 1898. The largest city in the United States, New York possessed unique technological, financial, commercial, and legal resources that were essential to his success as an inventor. He moved there in 1869 to work in the center of the telegraph industry and soon established his reputation as a talented electromechanic. He left after about eighteen months but only to cross the Hudson River to New Jersey, where his manufacturing ventures in **Newark, New Jersey**, were linked to the metropolis. Even rural **Menlo Park, New Jersey**, where he set up his **laboratory** in 1876, was within the city's economic and intellectual orbits—close enough to indulge his love of the **theater** on occasion. In fact, it was probably while creating his **electric light system** at Menlo Park that Edison drew most heavily on New York resources, including the financial backing of **J. P. Morgan**, the legal advice of **Grosvenor Lowrey**, and the publicity opportunities of the city's newspapers. Edison moved back to Manhattan in 1881 to oversee construction of the **Pearl Street Station** ("First District") and build up manufacturing capabilities at the **Edison Machine Works** and the **Electric Tube Company**. He and his family lived in a series of apartments in fashionable neighborhoods for five eventful years. In that time, he created a small **laboratory** at the **Bergmann & Company** factory, orchestrated the start of the commercial electric utility industry, and endured the death of his wife **Mary Stilwell Edison**. He moved out of the city for good in 1886, when he married **Mina Miller Edison** and settled in suburban **Llewellyn Park, New Jersey**. Even so, Edison kept deep ties to New York. The **Edison Electric Light Company**, successor **Edison General Electric Company**, the **North American Phonograph Company**, and the **Edison Ore Milling Company**—all based in Manhattan—were central to his activities for the rest of the century, even as he developed his manufacturing bases around **West Orange, New Jersey**, and Silver Lake, New Jersey, and personally spent large amounts of time at the Ogden ore milling plant in rural New Jersey. In the new century, the **National Phonograph**

Company had recording studios in Manhattan, and the **Edison Manufacturing Company** produced motion pictures there and in the Bronx (by then part of the consolidated city). Edison also relied on New York patent attorneys for the rest of his career.

NEWARK, NEW JERSEY. When Edison moved to Newark in 1870, the city was a fast-growing manufacturing center and a magnet for immigrants. Its proximity to **New York City**—Edison's recent home and the capital of American communications enterprises—gave him continued access to **Western Union, Gold and Stock**, and similar companies. Edison used his first contracts as an inventor to rent a small Newark shop to work on printing telegraphs and autographic telegraphy (a system for reproducing handwritten or other non-printed material). That shop became the Newark Telegraph Works and then Edison & Unger (with **William Unger** as a partner) in a larger building on Ward Street. There, Edison brought machinery and workmen from the **American Telegraph Works**, including **Charles Batchelor**. He also started a local news service (the News Reporting Telegraph Company) where he met the young **Mary Stilwell**, whom he married in December 1871.

As Edison focused less on manufacturing and more on designing instruments, he established his inventive reputation in New York and beyond with his work on **automatic telegraphs**, district telegraph systems, and the **duplex telegraph** and **quadruplex telegraph** systems. In 1872, he became partner with **Joseph Murray**, who helped with both the manufacturing and inventive work. Late the next year, he set up a small **laboratory** (above the Ward Street shop that he and Murray shared) and was joined there by **Charles Batchelor**. Determined to put his energy into inventing, Edison dissolved his manufacturing business with Murray in 1875 and created a larger independent laboratory in the building they had used. He took with him Batchelor, machinist **John Kruesi**, and **James Adams**; he soon hired his nephew **Charles Edison**. Secure in his ambition to make a living by inventing, Edison moved in 1876 with his family (which included his daughter

Edison's Ward Street shop in Newark (ca. 1872), where he set up his first laboratory. *Courtesy of Thomas Edison National Historical Park.*

Marion Estelle and son Thomas Jr., both born in Newark) to **Menlo Park, New Jersey**. Just six years later, having invented a complete **electric light and power system**, Edison turned again to manufacturing. Needing to expand production and lower the cost of his **electric lamp**, he approved moving his lamp factory from Menlo Park to East Newark (also called Harrison), just outside the Newark city limits. There, factory manager **Francis Upton** had a plant with room to grow and an abundance of cheap, semiskilled (largely female) labor in the Newark region. The large facility supplied lamps to the **Edison General Electric Company** and successor **General Electric Company** until 1929. Newark was also the longtime home of Edison rival **Edward Weston**.

NORTH AMERICAN PHONOGRAPH COMPANY. This firm was organized by **Jesse Lippincott** and incorporated in New Jersey in July 1888, though its principal offices were in **New York City**. Its purpose was to lease or sell **phonographs** and **graphophones** and their accessories (including recordings) through a network of licensed regional affiliates or "subcompanies." Lippincott assigned to it the patent rights he had acquired, including rights to the phonograph that he gained from buying a controlling interest in the **Edison Phonograph Company**. North American struggled from the start due to a weak market for the instruments, Edison's manufacturing problems, and Lippincott's highly controversial obligations to **Ezra Gilliland** and **John Tomlinson** related to his acquisition of Edison's patent rights. As a result, Lippincott was effectively bankrupt by May 1891. Because of debts owed to himself and the **Edison Phonograph Works**, Edison was able to take over North American's management in 1891. He put the company into receivership three years later and in 1896 purchased its assets and folded them into the **National Phonograph Company**.

O

OESER, MARION ESTELLE (EDISON) (1873–1965). Marion was the eldest child of Edison and his first wife, **Mary Stilwell Edison.** She was named for Edison's sister **Marion Edison Page**, but he gave her the telegraphic nickname "Dot." After her mother's death, Marion sometimes joined her father at work (he referring to her as "George") and on his travels, including the 1886 honeymoon trip with **Mina Miller Edison** (also joined by **Ezra Gilliland** and his wife). Like her siblings, Marion was never fully embraced by her stepmother, with whom—just eight years apart—she had a fraught relationship. Marion attended day schools in New York City and had a single unhappy year boarding in Massachusetts. She was traveling in Europe with a tutor at the end of 1889 when she contracted smallpox. Scarred physically by the disease and emotionally by the evident lack of concern from home, she remained abroad for two decades, except for a brief visit to the United States in 1892. She married a German military officer, Oscar Oeser, in 1895. Marion reunited with her father and stepmother (and the three half-siblings she scarcely knew) on their European tour in 1911. Her marriage, strained by wartime and Oscar's infidelity, ended in 1921. Marion returned to the United States and lived first in East Orange, New Jersey, and then in Connecticut, cultivating the love of opera she had acquired abroad. *See also* EDISON, THOMAS ALVA, JR.; EDISON, WILLIAM LESLIE.

OGDEN ORE MILLING PLANT. *See* NEW JERSEY AND PENNSYLVANIA CONCENTRATING WORKS (NJPCW).

ORE (GOLD) REFINING. Edison first expressed interest in detecting and processing low-grade gold ores in 1875 but did little at that time. His first substantial experiments took place four years later, when he tried to design a process for separating **platinum** from the black sand residue (tailings) at gold mines in hopes of securing a good supply of platinum for his experimental **electric lamps**. That project led in 1880 to the development of an electromagnetic iron-ore separator. Edison returned to gold ores in 1887, when he placed **William K. L. Dickson** in charge of research on the subject while focusing his own efforts on iron-ore separation. He picked up the subject again eleven years later amid his ongoing iron-ore work. He soon developed a dry placer process to separate gold from gravel in western mines where hydraulic processes could not be used. His dry process used screens to size pieces of gravel and gold; pieces of similar size went to a separator where a blast of air from a centrifugal fan separated the lighter gravel from the heavier gold. In 1900, Edison sent **John V. Miller** (his brother-in-law) and Cloyd Chapman to New Mexico to supervise construction of an experimental processing mill using his dry process. The amount of gold processed at the mine turned out to be too small to make a profit. Edison subsequently

Ogden plant for concentrating and purifying iron ore, 1893. *Courtesy of Thomas Edison National Historical Park.*

advertised the process to find other mine properties and sent Chapman to New Zealand at the behest of a group of investors, but he never succeeded in making the process a commercial success. *See also* EDISON ORE MILLING COMPANY; ORE (IRON) MILLING.

ORE (IRON) MILLING. The recovery of commercial-grade iron from overlooked lean ore deposits was one of the major projects of Edison's career. Initially an offshoot of his 1879 attempt to extract raw **platinum** from the waste sands of hydraulic gold mines, his efforts spanned the last twenty years of the nineteenth century; the project consumed most of his time and millions of his own dollars in its final decade.

Edison exploited iron's magnetic properties to separate it from the surrounding rock. His first separator was a simple machine. Fine ore fell by gravity from a hopper in a stream past a powerful electromagnet. The magnet diverted the metallic material, bending its path toward a bin below, while the nonmagnetic waste material ("gangue") fell straight into a waste bin. Designed to work lean ores that piled up unprofitably at mines, the sep-

arator could concentrate ore from as little as 15 percent iron into a product that was 65 percent iron (the degree necessary for making steel by the increasingly important Bessemer process). Edison patented this basic design in 1880; over the next two years, he successfully applied for four more patents, all of which became the property of the **Edison Ore Milling Company.**

In 1881, Edison attempted two field experiments on the iron-rich black sands of beaches on Long Island and in Rhode Island. The first was aborted after the sand washed out to sea; the second proved the operation of the separator, but the concentrate was too fine for commercial furnaces to handle. Edison abandoned the subject for several years after these disappointments. In the late 1880s, encouraged by laboratory experiments (notably by **William K. L. Dickson**), he established experimental plants in Humboldt, Michigan, and Bechtelsville, Pennsylvania, that he hoped would produce Bessemer-grade ore at economically justifiable costs. Neither did, though this was not entirely through the fault of the processes.

In 1890, Edison directed the **New Jersey and Pennsylvania Concentrating Works** to set up a plant near Ogdensburg, New Jersey, a facility that spurred years of intensive experimentation and innovation. Among the outcomes were separators for handling fine dusty ores and extracting harmful phosphorous, a new magnetic "dipping needle" for locating iron deposits, and a raft of processes and machines for conveying, crushing, sifting, and drying ores and for binding the particulate concentrate into "briquettes" for shipment and smelting. Faced with huge economic losses and the improbability of competing against new sources of cheap, rich ore from Minnesota's Mesabi region, Edison closed the Ogdensburg plant at the turn of the century. His hopes of rescuing depleted eastern mines—and the furnaces that depended on them—remained unfulfilled. Edison did, however, eventually receive about fifty U.S. patents for iron-ore mining and ore processing, some of which he adapted to revolutionizing the manufacture of cement. *See also* EDISON PORTLAND CEMENT COMPANY; ORE (GOLD) REFINING.

ORTON, WILLIAM (1826–1878). As president of **Western Union Telegraph Company**, Orton staunchly supported Edison's early inventive work. The young Orton had a varied career in education, publishing, and government, including as commissioner of internal revenue under President Andrew Johnson. In 1865, he left government service to become president of the United States Telegraph Company (USTC). When Western Union bought USTC the following year, Orton became Western Union's vice president. He succeeded to the presidency in July 1867 and held that post until his death. Western Union thrived under Orton, in part because he pursued new technologies to increase the volume and quality of its services. Observing Edison's work for the rival **Gold and Stock Telegraph Company**, Orton marked the young inventor as "probably the best electro-mechanician in the country" apart from his own superintendent **George Phelps** (Israel 1998, 49). He drew Edison into his orbit with Western Union's acquisition of Gold & Stock in May 1871. An 1873 verbal agreement allowed Edison to use Western Union facilities in **New York City** to test experimental telegraph apparatus in exchange for giving the company the first right of refusal on resulting patents. This agreement was the first of several under which Orton's company funded Edison's experiments in telegraphy (including the **quadruplex telegraph**) and on the telephone (his **carbon-button telephone transmitter**). The two men clashed on occasion: Orton was dismissive of Edison's work in **automatic telegraphy**, and in 1875, Western Union petitioned the New Jersey Court of Chancery to prevent Edison from selling the quadruplex to another party. Yet Orton was a friend as well. When Edison learned of his mentor's untimely death, he remarked that "if I get to love a man he dies right away. [**Marshall**] **Lefferts** went first, and now Orton's gone, too" (*TAEB* 4:238 n. 1).

OTT, JOHN F. (1850–1931). An expert model and instrument maker, Ott was born in Jersey City, New Jersey, and attended the Cooper Union Institute in **New York City**. Thomas Edison and **William Unger** (Ott's cousin) hired him in 1870, and he worked for Edison until poor health forced his retirement in 1920. Over that half century, he fashioned instruments and translated rough sketches into working models, two critical processes that facilitated a wide range of Edison's inventions. Ott was one of Edison's most trusted assistants, serving as chief experimental instrument maker and superintendent of experimenters at the **laboratory** in **West Orange, New Jersey**. Edison recognized Ott's conceptual contributions by permitting him to take out several patents in his own name. Ott was the father of Ludwig ("Louis") Ott, who did important chemical research for Edison from 1902 to 1926, and the older brother of Frederick Ott, also a close Edison associate and model maker for decades. "Fred," as he was known, worked on the **kinetoscope** and was the subject of the first motion picture copyrighted in the United States—a five-second film of him sneezing. John Ott died one day after Edison and was interred in the same cemetery in West Orange. (Edison's remains were later moved to his **Glenmont** home).

P

PAGE, MARION WALLACE (EDISON) (1829–1900). Edison's sister Marion was the first child of **Nancy Elliott Edison** and **Samuel Ogden Edison Jr.** She was born in Vienna, Ontario, and moved with the family to **Milan, Ohio**. She married a local farmer, Homer Page, two years after Edison's birth and remained in Milan the rest of her life. In her last years, she bought and renovated the house where Edison was born. Edison was closer with her than with his other siblings (except perhaps **William Pitt Edison**) and named his first child Marion Estelle in her honor.

PAINE, THOMAS (1737–1809). The writings of Paine, a British-born political philosopher and agitator, reflected Enlightenment ideals and helped inspire rebellion in Britain's colonies and revolution in France. Edison first felt the "revelation" of Paine's works via the library of his father (Israel 1998, 8). **Samuel Edison** was an advocate of religious free thought, which drew heavily on Paine's scientific deism and critique of organized religions. Similarly influenced by Paine, young Thomas took little interest in his mother's traditional religious beliefs. He later wrote that "my mother forced me to attend [her church]—my father gave me Paine's Age of Reason" (Israel 1998, 8). A lifelong admirer, Edison wrote an introduction to a collection of Paine's works published in 1925. *See also* RELIGION AND RELIGIOUS VIEWS.

PAINTER, URIAH HUNT (1837–1900). A longtime Washington correspondent for the *Philadelphia Inquirer*, Painter parlayed his close connections with Republican lawmakers into a career as a lobbyist and promoter. He met Edison in the early 1870s in connection with the **Automatic Telegraph Company**. Painter was among the founders and principal investors in the **Edison Speaking Phonograph Company** (ESPC), and in 1878, he posed with Edison, **Charles Batchelor**, and the **phonograph** for a photograph at Mathew Brady's studio in Washington, D.C. He also invested in the Edison lighting interests and used his Washington influence on their behalf. Painter's relationship with Edison soured in the late 1880s over formation of the **Edison Phonograph Company** and what Painter saw as an attempt to usurp rights of the ESPC and its stockholders.

PARIS INTERNATIONAL ELECTRICAL EXPOSITION (1881). Known in French as l'Exposition Internationale de l'Électricité, this event, from 10 August to mid-November 1881, featured advances in electrical technology. Edison's exhibit, supervised by **Charles Batchelor**, included an Edison motor driving a sewing machine, telegraphs, telephones, **stock tickers**, and the **electric pen**. But the focus was on Edison's incandescent **electric lighting system** (run by the "C" dynamo, the largest in the world). Presented to advantage by Batchelor and Otto Moses, the lighting display helped secure Edison's international reputation in the field. King Kalakaua of Hawaii and French prime minister Léon Gambetta were among the attendees who, according to Batchelor, "densely packed" the Edison rooms "every night" (*TAEB* 6:159). Moses noted "the complete success of our illumination. . . . I never saw a more beautiful sight" (*TAEB* 6:144). Edison's lamp was recognized as the most efficient, and Edison garnered the Exposition's highest award and appointment as an officer in the French National Order of the Legion of Honor. His rivals were less successful, with Batchelor and Moses gleefully chronicling **Joseph Swan's** "miserable attempt" (*TAEB* 6:142) at lighting, **Hiram Maxim's** "mess" of a display (*TAEB* 6:145), and a fire started by St. George Lane-Fox's lamp. Edison agents parlayed the exhibit's success into formation of lighting companies in Europe and contracts for lighting systems at the Paris Opera House, Milan's La Scala, and railway stations in Germany. The event also helped spur Edison's isolated lighting business in the United States and Europe. After the triumph in Paris, attorney **Grosvenor Lowrey** informed Edison, "Nobody for a moment questions now that you are the great man" (*TAEB* 6:224–25).

PARIS UNIVERSAL EXPOSITION (1878).

The third Paris world's fair, also known as the Exposition Universelle, ran from 1 May to 10 November 1878. A celebration of France's recovery from its defeat in the Franco-Prussian War of 1870–1871, the Exposition also showcased the latest in art, machinery, and technology, including the head of Frédéric Auguste Bartholdi's *Liberty Enlightening the World*, rubber tires, **Alexander Graham Bell's** telephone, and arc lighting. The Edison exhibit space, arranged by **George Bliss** and measuring twenty-five by twenty-five feet, featured the **phonograph**, the **carbon-button telephone transmitter**, and the megaphone. A 3 July demonstration before the Exposition's jury for telegraphic instruments featured messages recorded on an Edison clockwork phonograph in Paris and then replayed into an Edison carbon transmitter on a line to Versailles. Meanwhile, Herman Rothe sent messages from Versailles to Paris, where Edison associate **Theodore Puskás** recorded them by placing the telephone receiver against the phonograph mouthpiece. The jury members were not sufficiently impressed to bestow an award on Edison, but "the **Wizard of Menlo Park**" did garner the Grand Prize of the Exposition, prompting promoter George Beetle to note that Edison won the honor "not on any particular invention but as the Inventor of the age in which we live" (*TAEB* 4:488). Indeed, Edison was a hero to the citizens of the French Third Republic, a self-taught man and the antithesis of the aristocracy, one who nevertheless revolutionized sound. *See also* PARIS UNIVERSAL EXPOSITION (1889).

PARIS UNIVERSAL EXPOSITION (1889).

The Exposition Universelle, a world's fair, was held in Paris from 1 May to 6 November 1889. It was designed in part to celebrate the centennial of the French Revolution, though the organizers played down this aspect of the event to attract participation from European nations that were still monarchies. The exposition is most notable today for its central attraction—the Eiffel Tower—which has since become an iconic symbol of France. But for visitors to the fair, Edison's improved wax-cylinder **phonograph** was an even bigger draw. More than 4 million fairgoers listened to recordings on Edison's phonograph, exhibited at several venues.

The most important phonograph display was part of a larger exhibit of Edison's inventions, encompassing some 9,000 square feet in the enormous Palais des Machines. The Edison exhibit, which **William Hammer** organized, was the largest at the fair and included nearly the totality of Edison's inventive work to 1889. Some 493 of his inventions were on display, including those in telegraphy, telephony, underground tubing and wiring, batteries, lamps and lamp production, electric light and power, meters, and ore milling. The estimated cost was between $75,000 and $100,000 (about $2.75 to $3.1 million today).

Edison and his wife **Mina Miller Edison** attended the Exposition, touring the fair several times. They twice ascended to Gustav Eiffel's apartments at the very top of the tower. On the first occasion, they were received by Eiffel's sister and son-in-law and photographed by Count Giuseppe Primoli. On the second, Eiffel himself hosted, an event re-created today by mannequins of Eiffel and Edison inside Eiffel's apartment.

While most crowned heads of Europe declined to attend the Exposition, Edison was hailed as a symbol of republican virtue and achievement. The French received him with honors normally reserved for a head of state, including an official meeting with French president Sadi Carnot and numerous banquets in his honor. The government made him a commander of the Legion of Honour, and he received an appointment to the Order of the Crown of Italy.

Edison chafed at the formalities but used his time in France to meet with top scientists and engineers, including physiologist and photographer Étienne-Charles Marey, astronomer and photographer Jules Janssen, and chemist and microbiologist Louis Pasteur. Leaving Paris, Edison and Mina toured France, Belgium, Germany, and England. *See also* PENDER, JOHN; SIEMENS, WERNER VON.

PATENT SYSTEM. Edison received 1,093 U.S. patents and filed several hundred other unsuccessful applications; he also took out more than 1,200 patents in foreign countries. They were for him (like other inventors) the payoff for his efforts. Patents—a form of intellectual property—embody ideas. The American patent system, established under the U.S. Constitution, was designed to produce economic incentives for inventors. It granted a time-limited monopoly during which inventors could sell their rights to another party or make material goods based on the patent. The Patent Office administered the system, collecting fees from inventors to meet its operating expenses. Fees were quite modest in comparison with those of European countries, making U.S. patents accessible to many inventors.

To obtain a patent, inventors submitted a written application, measured drawings, and, until 1880, a small-scale model. The application required careful preparation (often by a lawyer) to describe the invention and delineate what the monopoly would cover. A Patent Office examiner scrutinized the application to determine whether the invention was new, useful, and unknown at the time of submission and whether the description was sufficient for a skilled worker to make the object. If approved,

a patent would be issued on payment of a final fee. If rejected, the inventor could amend the description or claims (though not the drawings or model) to make them more clear or precise. The cycle of rejection and amendment could go on for years before the patent was issued or the application abandoned.

A distinctive feature of the American patent system was its emphasis on rewarding the first person to conceive or make an invention. (European systems issued patents to the first person to file an application.) The American process put the burden on the inventor to demonstrate priority, and it was for this reason that Edison kept detailed records of his work. Disputes inevitably arose over competing claims to be the "first" with an invention. Two or more individuals might submit similar applications, or one might file for an invention already patented. In those cases, the Patent Office would declare an "interference," a quasi-judicial process that entailed a hearing before a special examiner. It often included depositions, testimony, exhibits, arguments, and, of course, lawyers. An applicant could appeal an unfavorable decision, first to the Patent Office commissioner and then to federal court in the District of Columbia. The Patent Office could not rescind a patent once issued, but it could grant another to a second person for the same invention. Patents could be challenged in federal courts for infringement (i.e., encroaching on the monopoly of another) or on grounds of priority, novelty, or ownership.

Inventors could also sue manufacturers or other inventors for infringing on their monopolistic rights. As his electric light patents became the subject of extensive litigation in the 1880s, Edison came to believe that patent laws favored infringers, who could continue to profit during the slow-course litigation. He argued for changing the law to give the first inventor the benefit of the doubt as a case went through the courts. Edison also objected to the fact that a U.S. patent would expire before the end of its normal seventeen-year term if a foreign patent on the same invention expired earlier. (Such had occurred with his basic **phonograph** patent, and the threat loomed over elements of his electric lighting system.) His growing frustration with the patent system even led Edison to suggest that the United States adopt the European system of "first to file."

As the head of a large research organization, Edison had to think carefully about what his assistants could patent. American law generally gave inventive credit to an employer unless the employee could be shown to have invented independently. For the most part, Edison's assistants understood that they were working on his ideas. Edison took out only a few joint patents with assistants because these could be challenged unless each coinventor could prove an equal contribution. Nonetheless, Edison encouraged assistants to take out ancillary patents on specific improvements while leaving the broader claims to him. By the twentieth century, employees in Edison's **laboratories**, factories, and operating companies created and patented many small but essential improvements to his innovations in sound recording, **batteries (storage)**, and cement.

Edison took out few patents in his last years, when his research focused on plants for producing **rubber**. Toward that end, he advocated for the Plant Patent Act of 1930 to enable breeders to patent new varieties of plants.

PEARL STREET STATION ("FIRST DISTRICT"). Edison built his first commercial central station in adjoining buildings at 255–257 Pearl Street on Manhattan's Lower East Side for the **Edison Electric Illuminating Company of New York**. The street served as a shorthand reference for both the station and the area it served (the "First District"), initially a roughly half-mile square of prominent offices, newspapers, brokerages, and banks (including **Drexel, Morgan & Company**). The station began service on 4 September 1882 after more than a year of construction. Its six dynamos, each driven directly by a steam engine, had a combined capacity for more than 7,000 lamps, though the district's buildings were wired (via eighteen miles of underground conductors) for

Dynamo room of the Pearl Street station in New York City, ca. 1882. Each of the six "Jumbo" generators was directly connected to its own steam engine. *Scientific American 47 (26 August 1882):* 127.

more than twice that number. To meet rising demand, the company added two dynamos in 1884 and built an annex station in 1886.

The Pearl Street plant ran almost without interruption until a major fire in January 1890. It was rebuilt rapidly and soon after converted to the **three-wire system** of distribution. The station operated until it was retired in 1894, well after the First District had lost its identity amid the electrification of Manhattan all around it. Reacting to the high cost of building the First District, Edison developed the **Village Plant** model and three-wire system of distribution for future stations. Only one other (in Milan, Italy) was built on the Pearl Street template.

PENDER, JOHN (1815–1896). Having amassed personal wealth as a British textile merchant, Pender organized and helped finance several major undersea cable companies in the 1860s and 1870s. These included the Eastern Telegraph Company, which rapidly became the largest cable company in the world. After Edison's 1873 trip to London to refine the **automatic telegraph**, Pender led a group of financial backers of the Edison system. The investment produced more frustration and discord than monetary return over the years, but Pender remained a friendly point of contact for Edison with Britain's business and technical communities. Pender also served as chairman of the Oriental Telephone Company, which promoted the telephones of Edison and **Alexander Graham Bell** in parts of Asia and Africa. Pender served in Parliament and was knighted in 1888. About that time, he began leading an effort to electrify portions of London with a combination of **alternating current** and **direct current** stations. Edison visited several of the new London stations while he and his wife stayed with Pender following the **Paris Universal Exposition (1889)**. Yet Pender would not commit to the Edison lighting system.

PHELPS, GEORGE MAY (1820–1888). As a respected inventor and superintendent of the **Western Union Telegraph Company**'s manufacturing shops in **New York City**, Phelps both rivaled and cooperated with Edison in the 1870s. After Edison entered the Western Union

orbit, president **William Orton** described him as "probably the best electro-mechanician in the country"—after Phelps (Israel 1998, 49). A native of upstate New York, Phelps had run his own machine shop and managed that of the American Telegraph Company before heading to Western Union. As Western Union's in-house inventor, he designed a printing telegraph and a **stock ticker** (both highly successful) and a magneto telephone. To Edison's consternation, the company evaluated the magneto telephone alongside his own carbon transmitter (eventually choosing the latter). Yet Phelps facilitated Edison's work too. His Western Union shop manufactured models for telegraph and telephone patent applications as well as the carbon telephone itself. He also periodically lent his facilities for Edison's use in experiments. Phelps retired from Western Union in 1884.

PHILADELPHIA INTERNATIONAL ELECTRICAL EXHIBITION (1884). The Franklin Institute organized this showcase of electrical apparatus that ran from 2 September to 11 October. It was the first such event in the United States after four in Europe (including the **Paris International Electrical Exposition (1881)**. Although international in name, most of the displays were American, and Edison's were the largest of all. He and his associated lighting companies mounted a comprehensive review of his career, featuring his electric light and power system, from lamps to the huge "Jumbo" central station dynamo. **William Hammer**, with exhibition experience from Paris, London, and Berlin, managed the Edison display. Its centerpiece was a tall column of 21,000 colored incandescent lamps, representing one day's output of the **Edison Lamp Company**. At the base, flashing lights spelled out the name "Edison." An Edison electrical plant provided much of the Exhibition's lighting through the new three-wire distribution system.

Edison, recently widowed, attended with his daughter Marion from 4 to 6 September and again from 16 to 19 September. He returned on 24 September for several more days. During these visits, he reconnected with his old friend **Ezra Gilliland**. The two men renewed their friendship and began planning collaborative projects, with consequences both good and bad for Edison.

The Exhibition closely followed a meeting of the British Association for the Advancement of Science in Montreal, and some of that group's members came to Philadelphia. Among them was **William Preece**, to whom Edison showed his **Edison Effect** lamp. It also overlapped with other professional gatherings: the National Conference of Electricians (to which Edison was appointed but did not attend), the **American Association for the Advancement of Science** (of which he was a member), and the inaugural meeting of the **American Institute of Electrical Engineers** (at which Edwin Houston read his paper about the Edison Effect lamp).

PHONOGRAPH. Edison burst into public consciousness in 1878 with this sound recording and playback machine. The phonograph became one of his signature inventions, closely identified with his daily work and reputation for decades. The machine took different forms over the years, but most fall into three broad categories: the tinfoil phonograph, the wax-cylinder phonograph, and the diamond disc phonograph.

Edison conceived the instrument in July 1877 while working on his **carbon-button telephone transmitter**. He thought that a point attached to a diaphragm moving over paraffin paper could emboss a physical record of sound into the paraffin. Finding that "the vibrations are indented nicely" into the paper, he imagined it possible "to store up & reproduce automatically at any future time the human voice perfectly" (*TAEB* 3:444). Edison had a design by November that machinist **John Kruesi** fabricated in December. The hand-cranked machine consisted of a metal cylinder on a shaft with a screw pitch; a sheet of tinfoil wrapped around the cylinder served as the recording medium. Edison tested it by reciting "Mary Had a Little Lamb," and the machine repeated his words to his amazed staff. Edison took it to the offices of *Scientific American*, where it elicited similar astonishment.

Edison and his tinfoil phonograph, 1878. *Courtesy of Thomas Edison National Historical Park.*

The tinfoil phonograph was an immediate sensation in the scientific and engineering communities and with the public. It made Edison's name a household word as "the **Wizard of Menlo Park**." But its commercial value was limited, as the **Edison Speaking Phonograph Company** quickly learned. The embossed tinfoil could not reproduce a full range of tones, nor was it durable. The phonograph sensation died down, exhibition crowds thinned, and Edison turned his energy to electric lighting in the latter part of 1878.

Edison renewed his efforts on the phonograph (his "baby") after **Alexander Graham Bell** and Charles Tainter introduced their own version—the **graphophone**—in 1886 and began selling it the next year. After several false starts, Edison had a "perfected" phonograph in June 1888. Like the graphophone, it recorded on a wax-covered cylinder that greatly improved both fidelity and durability. It also had an electric motor run by either a battery or an electric light circuit. Edison created several versions on this basic template (while also inventing a better wax material) over the next two years as the **Edison Phonograph Works** turned them out by the hundreds. He initially imagined using the phonograph for business correspondence both in stenography and for sending recorded letters by mail. Yet by 1889, he was producing musical recordings at the **laboratory**, hundreds of which he sent to the **Paris Universal Exposition**, where they drew huge crowds. By the next year, local phonograph companies in the United States were installing coin-in-slot phonographs in hotels, restaurants, and other public places where patrons could listen to the latest musical recordings.

Although Edison had imagined a flat-disc version of the original tinfoil phonograph in 1878, he did not develop a practical disc machine until 1911. His effort then was a response to the **Victor Talking Machine Company**, whose version of the **gramophone** surpassed Edison's **National Phonograph Company** in sales. Victor's discs were longer playing, more durable, cheaper, and easier to store than Edison cylinders. Victor also recorded more popular musicians and singers. Edison had resisted the idea of a disc phonograph because he thought cylinder recordings were of higher fidelity. He instead improved

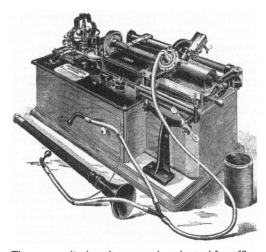

The wax-cylinder phonograph, adapted for office dictation (1893). The ear tubes and speaking horn rest in front of the machine; a cylinder is at right. *Courtesy of Thomas Edison National Historical Park.*

the cylinder phonograph and recordings, introducing the Amberola phonograph and Blue Amberol cylinders (made of a hard celluloid compound) in 1908. But this effort came too late to stave off the challenge of Victor's disc machine.

Development of the Edison Diamond Disc Phonograph (introduced in 1911) was a team effort. **Jonas Aylesworth** developed a long-lasting new resin for the discs. Edison concentrated on the recorder and reproducer, switching from a sapphire stylus to the more durable diamond and introducing a floating-weight design for the recorder/reproducer arm. The engineering department took the lead on the instrument's mechanical design.

As with the wax-cylinder phonograph, Edison focused on maximizing fidelity, for which he believed consumers would pay a premium. He also thought the market would reward the quality of the recordings rather than performers' reputations, and he insisted on personally choosing the musicians for Diamond Disc recordings. Edison's marketing campaign reflected these convictions. In 1915, he instituted a series of "Tone Tests" in which an artist would perform live on a darkened stage or behind a curtain as a phonograph played a recording of the same artist and composition. At some point, the artist would stop, leaving the audience unable to tell if the ongoing music was live or recorded. Although the Edison Diamond Disc Phonograph was technically superior to the Victor machine, Edison's strict control over the recording catalog placed his instrument at a commercial disadvantage. His company never overtook Victor. *See also* TALKING DOLL (TOY PHONOGRAPH).

PHONOPLEX TELEGRAPH. Edison created this form of multiple telegraphy in 1885 to overcome a limitation of other multiple telegraph systems, such as the **duplex telegraph** and the **quadruplex telegraph**, which could work only between the endpoints of a line. Edison specifically intended his new method to serve intermediate stations. The phonoplex had roots in his **railway ("grasshopper") telegraph**, whose rapidly vibrating signals did not register on conventional Morse instruments.

Edison adapted a telephone receiver, already sensitive to the oscillations of the human voice, to respond to such signals. The result was the creation of two independent channels in a single wire, one for terminal stations and the other for intermediate stations. Other inventors had worked along similar lines in Belgium and Great Britain, but Edison's phonoplex took hold only in the United States. It became one of his more commercially successful inventions, with more than a dozen railroads (including the giant Pennsylvania Railroad) and the **Western Union Telegraph Company** subscribing to it as of 1901.

PHRENOLOGY. This was a widespread nineteenth-century pseudoscience in which the shape of an individual's skull and its bumps were said to indicate mental traits and character. Edison was familiar with phrenology by 1878. In February of that year, the *Phrenological Journal* published a portrait and biography of the inventor, including a discussion of the mental and personal characteristics purportedly evident in his portrait. Later that year, Edison also sat for a phrenological bust commissioned by the Phrenological Institute of New York. Edison personally believed that phrenological study of the shape of a head and its features provided useful information but that the bumps did not. For that reason, he sometimes asked prospective employees to submit a photograph or commented on their features after seeing them.

PLATINUM. This noble metal was so important to Edison's development of the incandescent **electric lamp** that his search for a large supply cemented his popular reputation as the **Wizard of Menlo Park**. Platinum is hard, electrically conductive, and nonreactive and has a high melting point. It was used in telegraph instruments, where Edison would have learned something of its properties; guided by his interest in **chemistry**, he wrote a paper on the metal and its uses that appeared in 1874. As a fledgling inventor, Edison chose platinum (often calling it by an old name, "platina") for delicate contact surfaces in printing telegraphs, telephone transmitters, and the **electro-**

motograph receiver. When he seriously took up the creation of an electric lamp in 1878, he planned to make incandescent "burners" and thermal regulators from platinum-alloy wire that could endure intense heat without melting or oxidizing. Confident in his platinum-spiral lamp in spring 1879, Edison began seeking supplies of the costly metal (costly due to scarcity and the difficulty of refining). He made personal inquiries, dispatched an assistant to Quebec and California (where platinum was found with gold), and searched state geological surveys. He also sent 2,000 **electric pen** copies of a handwritten inquiry to postal and other authorities in mining regions, mainly in the American West—an effort that inspired the newspaper cartoon *The Wizard's Search*. Always attuned to the properties of materials, Edison created a vacuum process (described in an 1879 paper for the **American Association for the Advancement of Science**) that enabled thin platinum wire to survive ever higher temperatures. After replacing the metallic "burner" with the carbon filament, he ended up not needing as much platinum as anticipated (though he still used it for the lead-in wires entering the glass globe). But his connections to western mining inspired a new research project: gold **ore refining**.

POLYFORM. Drawing on the wide variety of chemicals and drugs stocked in his **laboratory** in **Menlo Park, New Jersey**, Edison formulated a topical analgesic in early 1878 that he promised "would stop any kind of pain immediately"—including, perhaps, his own neuralgia (*TAEB* 4:228 n. 3). Its composition varied but generally included alcohol, chloroform, ether, morphine, camphor, chloral hydrate, and occasionally opium, among other readily available ingredients. This multi-ingredient formula, combined with Edison's penchant for using Greek roots in naming his inventions, led Edison to name his creation "polyform." An interview with the *New York World* in March 1878 helped create a buzz around the new concoction, and Edison was besieged by requests for either the formula (which he gave away freely) or the distribution rights. In 1879, Edison assigned his pending U.S. patent application to mining associate Charles Lewis and several partners "to rid myself of the awful amount of correspondence & trouble which the newspaper publications caused" (*TAEB* 5:795 n. 4). Lewis's group formed the Menlo Park Manufacturing Company in **New York City**, promoting "Edison's polyform" as a treatment for neuralgia, rheumatism, and headaches. The company later sought to include Edison's autograph and portrait in the copyright on its labels, but Edison refused. In 1890, when Lewis tried to form a new company called the Edison Polyform Company, Edison instructed his secretary to "write Lewis that if any further attempts are made to bring the polyform out that I shall Knock it in the head" (*TAEB* 5:795 n. 4). Edison never received a patent on polyform in the United States, though he did in Great Britain under the different **patent system** there.

POPE, EDISON & COMPANY. *See* POPE, FRANKLIN LEONARD.

POPE, FRANKLIN LEONARD (1840–1895). Pope was a noted electrical engineer, inventor, author, and patent attorney. He first encountered Edison as editor of *The Telegrapher* (August 1867–February 1868), when he published articles submitted by Edison. In the spring of 1869, he assisted with Edison's experiments on a double transmitter. When Pope left that summer on a trip to the Great Lakes, Edison replaced him as superintendent of Samuel Laws's Gold and Stock Reporting Telegraph Company and redesigned the company's stock printers. **Gold & Stock Telegraph Company** took over the Laws company in September 1869, and Pope and Edison collaborated on an improved printer; Edison also started boarding at the Pope family home in Elizabeth, New Jersey. By October 1869, they had formed Pope, Edison & Company with James Ashley, Pope's successor at *The Telegrapher*, as a silent partner. To market the improved Pope and Edison printer, they launched the Financial and Commercial Telegraph Company, which was soon acquired— along with their patents—by Gold & Stock. However, the partners retained patent rights

for private lines, and with **Marshall Lefferts**, president of Gold & Stock, they established the American Printing Telegraph Company to serve that market. Gold & Stock took over that company as well in 1871. By then, Edison's relations with both Pope and Ashley had deteriorated. Ashley subsequently attacked Edison in the pages of *The Telegrapher*, and while Pope continued to be outwardly cordial, his relationship with Edison was never the same. Pope would later challenge Edison's claim as the inventor of the carbon-filament **electric lamp** and align himself with **George Westinghouse** and **alternating current (AC)**. He died of accidental electrocution by AC wires at his home.

PORT HURON, MICHIGAN. Port Huron was a busy port town, where water and trade funneled into the St. Clair River between Lake Huron and Lake Erie, when the young Edison's family moved there in 1854. Father **Samuel Edison** joined the burg's thriving lumber business, and the boy received his limited formal **education** there until mother **Nancy Edison** took over his tutelage. The family home was a two-story house on the north side of Port Huron built by the first purveyor of the Fort Gratiot military reservation. Samuel erected a 100-foot observation tower nearby that became a tourist attraction for its views of Lake Huron and the St. Clair River. He also operated a truck garden. Young Edison helped with the latter enterprise until the Grand Trunk Railway opened a line from Port Huron to Detroit (about sixty miles away) in 1859, when he started working as a newsboy aboard the passenger trains. The experience introduced him to telegraphy, the newspaper business, entrepreneurship, and the library of the Young Men's Society in Detroit, where he spent hours reading between outbound and homeward trips. He also took the opportunity to set up a rolling chemical laboratory in a baggage car. Edison got his first job as a telegraph operator in Port Huron in 1862. His departure the next year for an operator's job on the railroad at Stratford Junction, Ontario, marked the start of his career as an itinerant telegrapher. *See also* CIVIL WAR TELEGRAPHY.

PORTER, CHARLES T. (1826–1910). A lawyer by training, Porter turned to steam engine design and, in 1862, applied a novel governor device to create the first commercially successful high-speed stationary steam engine. Edison and his assistants designed their large direct-connected, central-station dynamo around Porter's engines and collaborated with him on improvements. Edison selected Porter engines for the early central stations at London's **Holborn Viaduct**, at **New York City**'s **Pearl Street Station ("First District")**, and in Milan, Italy. Porter was less successful as a manufacturer and was ousted from his own company (the Southwark Foundry in Philadelphia), but Edison continued to work with him. Edison agreed to build and market Porter's engines and allowed him to experiment at the **Edison Machine Works** and the **laboratory** in **West Orange, New Jersey**, in hopes of greatly improving engine efficiency. Although Porter obtained several patents, he did not develop an engine that Edison could make or sell. He fell into ill health, and in 1891, Edison ended financial and experimental assistance to his project.

PREECE, SIR WILLIAM HENRY (1834–1913). Preece was a regional electrical superintendent for the national telegraph system of the British Post Office when he met Edison in 1877—affording the young inventor an entrée to the British technical community and, potentially, favorable consideration by postal authorities. Their introduction occurred in the context of Edison's work for **Western Union Telegraph**, but Preece's several visits to Edison's **laboratory** in **Menlo Park, New Jersey**, kindled a collegial friendship. Edison sent an early tinfoil **phonograph** to London, and Preece gave the first phonograph exhibition outside the United States (using a replica instrument) in February 1878. Around that time, Preece advanced to chief electrician at the Post Office, giving him more sway in decisions about telegraph systems and the telephone. Edison thought of Preece as his agent for the telephone in Great Britain and was incensed when he came to believe, in May 1878, that Preece had colluded with countryman David Hughes to take

credit for the variable-resistance carbon telephone transmitter and microphone. Edison aired his grievances in public and dragged scientific luminaries like **William Thomson** into the **microphone controversy**. The rift began to heal in 1881, when Preece, by then a fellow of the Royal Society, endorsed the Edison lighting system. The two men served on a committee formed in 1884 to establish standards for measuring the intensity of light. With friendly relations restored, Edison showed Preece some peculiar phenomena in his lamps, and it was Preece who, describing what he had seen, coined the term **Edison Effect**. In the late 1880s, Preece feuded bitterly with electrical engineer Oliver Heaviside over the latter's counterintuitive suggestion that adding inductance on telephone and telegraph lines would increase the clarity of signals. (Heaviside's idea greatly improved long-distance telephone service.) Preece became the Post Office's chief engineer in 1892 and retired in 1899.

PRESCOTT, GEORGE BARTLETT (1830–1894). An important technical authority on the telegraph, Prescott played key roles in Edison's early inventive career. Born in Kingston, New Hampshire, Prescott began working as an operator in 1847. He served as chief operator, manager, and superintendent for several firms and published *History, Theory, and Practice of the Electric Telegraph* (1860). Named chief electrician of **Western Union Telegraph Company** in 1870 by **William Orton**, Prescott's responsibilities encompassed improvements to the company's lines and evaluating new inventions. Prescott and Edison joined forces four years later to develop and market the latter's designs for multiple telegraphy (particularly the **quadruplex telegraph**), with profits to be divided equally between them. However, soon afterward, a cash-poor Edison sold his quadruplex patent rights to Western Union rival **Jay Gould**, prompting extensive litigation and his estrangement from Prescott. As the quadruplex court fights dragged on, Edison and Prescott did manage to work together on other matters, including the quadruplex business in Europe. Edison contributed to Prescott's *Speaking Telephone, Electric Light,* and Other Recent Electrical Inventions (1879), and Prescott tested Edison's **carbon-button telephone transmitter** over Western Union lines in 1878. Prescott produced other books on electrical and acoustic inventions before retiring from Western Union in 1882. He published the comprehensive *Dynamo-Electricity: Its Generation, Application, Transmission, Storage and Measurement* two years later.

PRESS RELATIONS. The *Scientific American* announcement about the **phonograph** in December 1877 generated a wave of attention to Edison in the daily and periodical press. By the end of March 1878, reports and interviews appeared frequently in **New York City** newspapers, and reporters traveled to **Menlo Park, New Jersey**, from other cities to talk with Edison and write about his invention. Newspapers across the United States and Europe reprinted these articles. As a former press-wire telegraph operator, Edison was familiar with newspaper practices and made friends with reporters and editors, some of whom had been operators themselves. He cultivated good relations with the press for the rest of his career and used the coverage to promote himself and his inventions. An engaging raconteur, Edison became a favorite subject for some journalists who returned time and again to interview him. Edison was one of the most famous people in the world when he died, and most newspapers at home and abroad marked the event with banner headlines and retrospective accounts of his life and career. *See also* MICROPHONE CONTROVERSY; WIZARD OF MENLO PARK.

PUSKÁS, THEODORE (TIVADAR) (1844–1893). Born to a prominent Transylvanian family in Pest, Hungary, Puskás managed a railway in his home country and a silver mine in Colorado before meeting Edison in December 1877. Edison had just demonstrated the **phonograph**, and Puskás began negotiating for the rights to it and the **carbon-button telephone transmitter**. Edison took an immediate liking to the Hungarian (perhaps due to his "displaying a handfull of $1,000 Bills") and appointed him to represent the phonograph and telephone interests in continental Europe (*TAEB* 3:677). Puskás agreed to pay the large expenses of patents and to negotiate the sale of patent rights in return for part of the proceeds. Puskás first demonstrated the phonograph in London, but his March 1878 presentation to the French Academy of Sciences announced the instrument on the continent to "great excitement and a storm of applause" (*TAEB* 4:167). Edison recognized Puskás as one of the few telecommunications promoters who "fully understand[s] the conditions over there" (*TAEB* 4:491). Yet the efforts of Puskás on behalf of Edison's telephone produced mixed results, and Edison eventually revoked his power of attorney. Puskás also hoped for a large role in the European electric light business. However, unable to advance as much money as Edison wanted for experiments, he ceded the lead to other investors, though he did help organize the Edison Electric Company of Europe (based in **New York City**) and the **Compagnie Continentale Edison** in Paris. He also brought in a young **Nikola Tesla** to work for Edison lighting interests in France. Puskás died in Budapest at age forty-eight. *See also* PARIS UNIVERSAL EXPOSITION (1878).

PYROMAGNETIC GENERATOR AND MOTOR. These complementary devices were offshoots of Edison's interest in the **direct conversion** of coal into electricity. He designed them to convert heat energy directly into either electricity (in the generator) or mechanical motion (in the motor) without the use of dynamos, steam engines, or other intermediate machinery. Both were based on the familiar idea that the magnetization of metals can change dramatically with temperature. In an 1887 paper for the **American Association for the Advancement of Science**, Edison explained that because changes in a magnetic field will induce an electric current (and vice versa), "it occurred to me that by placing an iron core in a magnetic circuit and by varying the magnetizability of that core by varying its temperature, it would be possible to generate a current in a coil of wires surrounding this core." The effect would be like that when an armature winding experiences a changing magnetic field, except that there was no need of rotation (in the generator) or an energizing current (in the motor). The idea struck Edison on vacation in 1886, and he returned to it at his Florida winter home and **laboratory** the following year. He drafted patent applications on the motor in March and May and one for the generator in May 1887. Prototypes made that summer proved the concept but were not practically useful. At least two other inventors, notably **Nikola Tesla**, explored similar ideas somewhat before Edison.

Q

QUADRUPLEX TELEGRAPH. In 1874, Edison designed the first practical telegraph for simultaneously transmitting four messages (two in each direction) over a single wire. He did it by combining the standard **duplex telegraph** (two messages in opposite directions) with a diplex system (two messages in the same direction). Unlike other inventors working on diplex systems who used weak and strong batteries to produce two signals of different strengths, Edison combined two types of receivers: one with a polarized relay that responded to changes in the polarity of the current and the other with a standard neutral relay. Getting this design to work proved difficult. Current reversals created a momentary drop in current strength that caused the standard relay to lose its magnetism just as it was supposed to respond to the other signal. Edison solved this problem with what he called a "bug trap," which isolated the effect electromechanically so it would not interfere with the signal.

As he worked on his quadruplex, Edison turned to **George Prescott**, chief electrician of **Western Union Telegraph Company**, for help in getting access to Western Union facilities for his experiments. In return, Edison agreed to name Prescott as coinventor and divide any profits equally. Western Union president **William Orton** approved the arrangement, and in June 1874, Edison began experimenting at Western Union's headquarters with instruments made for him by the company's factory. By the end of September, Edison had the new system running between **New York City** and **Boston**, and Western Union soon placed it into commercial service on that line. Because the quadruplex had the potential to greatly reduce the number of wires the company needed to expand its service throughout the United States, Orton considered it to be "an invention more wonderful than the Duplex" (Israel 1998, 99).

Negotiations among Edison, Prescott, and Orton over quadruplex patent rights stalled in December 1874 at the very moment that the economic downturn caused by the Panic of 1873 threatened Edison with bankruptcy. With Orton out of town, Edison began negotiating with financier **Jay Gould**, who hoped to use both the quadruplex and Edison's **automatic telegraph system** on the lines of his **Atlantic and Pacific Telegraph Company** to challenge Western Union. By the time Orton returned in mid-January 1875, Gould had acquired Edison's rights to both systems. Western Union soon acquired Prescott's patent rights, and the two companies waged an intense legal struggle over the quadruplex that was ended only when Western Union acquired Atlantic and Pacific in 1877.

R

RAILWAY ("GRASSHOPPER") TELEGRAPH. This system was a means of signaling to and from moving trains, the obvious utility of which attracted several inventors by the mid-1880s. **Ezra Gilliland** co-owned a patent for a railroad telephone that he and Edison began adapting to telegraphy in late 1884. Edison took over the project and created a transmitter with two distinctive elements. One was a vibrating reed to open and close a switch hundreds of times per second. The other was an induction coil through which the vibrating signals passed to produce high-voltage pulses strong enough to induce corresponding signals in a telegraph wire along the tracks. The long "jump" from train to wire suggested the "grasshopper" nickname. It also led Edison to see the process as something other than electromagnetic induction. Rather, he thought it was akin to the action inside a condenser or perhaps indicative of a new form of energy. The system had a successful public test in February 1886, but the enterprise formed to develop it (the Railway Telegraph and Telephone Company) attracted few subscribers.

RANDOLPH, JOHN F. (1863–1908). A close associate of Edison in a variety of positions, Randolph started his career in 1878 as an office assistant at **Menlo Park, New Jersey**, and ended as private secretary to the inventor and his wife **Mina Miller Edison**. He kept Edison's personal and laboratory accounts for much of the 1880s and 1890s. In that time, he also became involved in the administration of Edison enterprises, serving eventually as sec-retary or treasurer of the **Edison Phonograph Company**, the **National Phonograph Company**, the **Edison Manufacturing Company**, the **Edison Storage Battery Company**, and the **Edison Portland Cement Company**. In 1894, he succeeded **Alfred Tate** as Edison's private secretary, a role that came to involve handling requests for money from Edison's two adult sons (**Tom Jr.** and **William Edison**). After Randolph died by suicide, Edison recalled him as "one of my best friends" and "one of the men I most highly prized" for his character and business acumen("John F. Randolph, Treasurer of All the Edison Companies, Puts End to Life by Shooting," *Newark Advertiser*, 17 February 1908).

REIFF, JOSIAH CUSTER (1838–1911). Reiff was a railroad financier who provided capital for Edison's work in **automatic telegraphy** at a crucial point in the young inventor's career, and the two men remained friends until Reiff's death. He was a protégé of William Palmer, under whom he served in a Pennsylvania cavalry regiment in the Civil War, acquiring the nickname "Colonel" (though not the rank). He joined Palmer as an investor in the **Automatic Telegraph Company** in 1870 and was the company's secretary. Reiff claimed to have put up as much as $175,000 in support of Edison's development work and salary from the company. He also loaned money for the inventor's manufacturing partnership with **Joseph Murray** and for Edison's investment in the street railway of his brother **William Pitt Edison**. When the **Edison Speaking Phonograph Com-**

pany reorganized in 1878, Reiff became an investor and director. Their roles reversed, however, and a year later, it was Reiff who sought a loan from Edison. He borrowed as much as $60,000, much of it not repaid when he died.

RELIGION AND RELIGIOUS VIEWS. Edison was a freethinker who expressed skepticism toward traditional religion, arguing that the truths of nature—uncovered through scientific investigation—should replace clerical authority and biblical myths. In this conviction, he followed his father, who "gave me [**Thomas**] **Paine**'s Age of Reason," rather than his mother, who "forced me to attend" her church (Israel 1998, 8). His rationalism also stood out from the mainstream Christian beliefs of his wife **Mina Miller Edison** and her devout parents. What he found valuable about organized religion were moral teachings like the Golden Rule, which he believed harmonized with nature.

Edison's views became controversial after a 1910 interview in which he seemed to deny the existence of God. As he subsequently explained, he believed not in a personal God but in a pervasive supreme intelligence that created the immutable laws of nature. Edison also had unconventional views on the immortality of the soul. He seems to have believed that consciousness resides in fundamental units or animate atoms that combine to make up each human being and that these exist prior to and after an individual's life. He even speculated about communicating with these "life units" and said in the early 1920s that he was working on a device to enable these entities to manifest themselves. However, in interviews, he gave conflicting statements about communicating with the dead, claiming at times that he was just giving reporters a good story and was not serious about such a device. *See also* MATTER (EDISON'S CONCEPTIONS OF).

ROWLAND, HENRY AUGUSTUS (1848–1901). A preeminent physicist, Rowland was the inaugural professor in that department of the research-oriented Johns Hopkins University from 1876 until his death. He and

George Barker visited Edison's **laboratory** in **Menlo Park, New Jersey**, in March 1880 to measure the efficiency of the Edison incandescent **electric lamp** at converting electrical energy into light. Their results, published in the *American Journal of Science*, supported Edison's claims. Thereafter, Edison confided in Rowland regarding Barker's and **Henry Morton**'s treatment of him, while Rowland sought Edison's help in obtaining a U.S. patent on a dynamo design that might fund a new laboratory at Johns Hopkins. (The case was among those lost by Edison attorney **Zenas Wilber**.) In 1883, Rowland used the annual meeting of the **American Association for the Advancement of Science** to issue a public "Plea for Pure Science" that, without naming Edison, contrasted the public enthusiasm for his practical successes with a corresponding lack of support for basic research. Their paths crossed again in 1889, when Rowland consulted on a proposal to generate hydroelectric power at Niagara Falls.

RUBBER RESEARCH. Having experienced chemical shortages from German suppliers during World War I, American manufacturers (especially in the rapidly expanding automobile industry) worried about depending on foreign supplies of rubber, which were controlled largely by British companies. Among them were Edison's friends **Henry Ford** and **Harvey Firestone**, who needed plentiful supplies of cheap rubber for automobile tires. Both men had tried unsuccessfully to develop overseas plantations and had experimented with growing rubber trees at winter homes in Florida. Edison himself considered building a factory to make the rubber he needed for **batteries (storage)** and disc record production. With rubber prices rising, the three men agreed to form the Edison Botanic Research Corporation in 1927, with Ford and Firestone funding most of the experimental costs. Edison focused on finding a plant that could be cultivated on American soil, then harvested and processed to yield sufficient latex at a cost competitive with imported rubber.

Edison devoted nine acres of his **Fort Myers, Florida**, estate to growing, harvesting, and processing plant varieties. He also received more than 17,000 plant specimens from hired field men, Union Pacific Railroad agents, private collectors, the New York Botanic Gardens, and the Arnold Arboretum in **Boston**. By 1930, Edison had identified goldenrod as a promising source because it grew quickly and could "be mowed like wheat and the rubber obtained by chemical solvents with very little machinery" (Israel 1998, 459). Edison's declining health prevented him from completing the project, and the rubber supply problem was instead solved by chemists who created synthetic rubber.

S

SAWYER, WILLIAM EDWARD (1850–1883).
The work of Sawyer, a journalist from Maine who became a prolific inventor, created one of the most serious challenges to Edison's legal claims to the invention of a practical incandescent **electric lamp**. Sawyer first crossed paths with Edison in 1875, when he charged Patent Office examiner **Zenas Wilber** with improperly allowing Edison to amend an application, setting off a sharply personal feud. Sawyer would publicly disparage Edison's work on the **telegraph**, telephone, and **phonograph**, but it was in electric lighting that the rivalry had real consequences. Sawyer and business partner Albon Man did make an incandescent lamp about the same time as Edison, albeit one with such low electrical resistance as to be impractical. When Edison revealed his own lamp in 1879, Sawyer and Man attacked it in the press as unoriginal, trading charges in print with Edison and **Francis Upton**. A patent infringement case ensued over the carbonized-paper filament. Sawyer and Man won an initial victory, though Sawyer, ravaged by alcoholism, died before the 1883 decision. Edison contested the ruling (on principle, he said), though he had long since stopped using carbonized paper. The Sawyer–Man patent was eventually issued and, after passing through several hands, came under the control of the **Westinghouse Electric Company**. In 1885, the Edison Electric Light Company sued Westinghouse in federal court, alleging that its use of Sawyer–Man patents infringed Edison's fundamental lamp patent (U.S. Patent 223,898). The 1891 decision in the case was a clear victory for the Edison interests and an affirmation of Edison's inventive priority on the lamp's defining feature: its slender, high-resistance carbon filament. The ruling was upheld by an appeals court in 1892 and subsequently by the U.S. Supreme Court.

SCIENCE (EDISON AND). Edison considered himself not a pure scientist but rather a "scientific inventor" (Albion 2011, 69). His approach to invention drew extensively on scientific knowledge and research methods but was focused on commercial invention rather than expanding human understanding of the natural world. Nonetheless, his research sometimes verged into the scientific as he explored little-understood phenomena and even proposed general theories to account for them. Because his inventive work was often at the cutting edge of scientific knowledge, it was of great interest to prominent American and European scientists. Edison was welcomed at meetings of major scientific societies, including the **National Academy of Sciences** and the **American Association for the Advancement of Science**. When he provided funding for the journal *Science*, he saw it as a way to contribute to the larger scientific community. Edison was also a proponent of engineering education because he appreciated the value of scientific training for engineers and recognized its importance for deriving design principles. However, in the twentieth century, as the rise of research universities helped to advance new theoretical approaches and industrial research and engineering came to emulate academic science, the scientific community no longer regarded Edison so warmly. *See also* ASTRONOMY; BARKER, GEORGE; CHEMISTRY; CROOKES, SIR WILLIAM; EDISON EFFECT; ELECTROMAGNETISM; FARADAY, MICHAEL; HELMHOLTZ, HERMANN VON; HENRY, JOSEPH; LABORATORIES; MATTER (EDISON'S CONCEPTIONS OF); MAXWELL, JAMES CLERK; MICROPHONE CONTROVERSY; MORTON, HENRY; TASIMETER; THOMSON, SIR WILLIAM (LORD KELVIN); THURSTON, ROBERT HENRY; UNKNOWN NATURAL FORCES; X-RAYS.

SCIENCE (JOURNAL). After consulting with Edison in early 1880, John Michels, a former editorial writer for the *New York Times* and contributor to scientific publications, boasted that he would soon publish "the leading Scientific Journal in this Country, and the only one which will be read abroad" (*TAEB* 5:721). The fruit of his efforts was *Science, a Weekly Record of Scientific Progress*. The journal focused on the United States and consisted of papers delivered before scientific institutions, occasional original papers, and reports on inventions and other items of scientific or technical interest. Edison was at first the journal's main benefactor. He advised Michels, bought ownership shares, leased a **New York City** office, advanced hundreds of dollars, and covered all expenses, including salaries for Michels and an assistant. Edison had little editorial input but made occasional critiques, as when he scolded Michels for publishing a paper on electric motors by "an idiot" whose musings were "nothing more nor less than a disgrace to a scientific paper" (*TAEB* 5:879). In February 1881, Edison decided to stop underwriting the journal because he was "to[o] busy to give it any attention"; he made a final payment a year later (*TAEB* 5:987). In April 1882, Michels asked Edison to take stock in a new company to run *Science*, but Edison decided "not have anything further to do with [its] publication" (*TAEB* 6:230 n. 2). *Science* suspended publication in mid-1882 but resumed in February 1883 under the patronage of **Alexander Graham Bell** and **Gardiner Hubbard**. The journal passed through other hands before the **American Association for the Advancement of Science** (AAAS) purchased the title in 1894. *Science* became the official publication of the AAAS in 1900 and had an estimated weekly readership of 570,400 in 2014.

SCIENCE FICTION. Edison freely speculated about the technological future, and his inventions and reputation inspired others to imagine how science and technology might change society. Edison-centered fiction proliferated amid the general astonishment that greeted the **phonograph**, and some of the first efforts played humorously with the inventor's snowballing fame. *New York Daily Graphic* reporter William Croffut was the first in 1878, penning an April Fool's story of Edison creating a machine to make food out of "air, water and common earth" (*TAED* MBSB10470X). Later that year, the satirical magazine *Punch's Almanack* published a series of humorous drawings depicting fanciful Edison inventions, such as the "telephonoscope" (for transmitting pictures and sound like an early television) and "anti-gravitation underclothing" (*Punch*, 9 December 1878). A more serious literary effort appeared that same year in the form of a short story by French author Auguste Villiers de l'Isle-Adam about a female android created by Edison. Villiers de l'Isle-Adam expanded the story into an 1886 novel titled *L'Eve future* (*Tomorrow's Eve*), in the preface to which he explained how the publicity surrounding Edison had generated a legend that now belonged "to the world of literature" (Villiers de l'Isle-Adam 1982, 3).

In the 1890s, Edison inspired a series of dime novels depicting the adventures of Tom Edison Jr. Although the protagonist of these stories was an orphan, both his name and his ingenuity connected him to the famous inventor. Stories about boy inventors predated Edison, but scholars have given the name "Edisonades" to the genre, the best-known examples of which are the Tom Swift novels.

Edison himself was the subject of an 1898 sequel to H. G. Wells's *The War of the Worlds*. British and American magazines had serialized the original the year before, and unauthorized versions appeared in American newspapers, such as the *New York Evening Journal*. Arthur

Brisbane, editor of the *Journal*, commissioned a sequel from George Serviss, an author and lecturer known for his work on astronomy who later achieved recognition for his science fiction stories. The 1898 sequel, titled *Edison's Conquest of Mars*, had Edison develop weapons and other technologies (including spaceships, an antigravity machine, and a disintegration ray gun) that enable humans to launch a counterinvasion and defeat the Martians. Edison complained that the publication seemed to present him as Serviss's collaborator, and he tried to distance himself from it.

Edison's concerns may have arisen from his unrewarding collaboration on another piece of speculative fiction. In 1890, journalist **George Parsons Lathrop**, inspired by Edward Bellamy's popular novel *Looking Backward 2000–1887* and his own interviews with Edison, proposed that they cooperate on a futuristic novel to be called *Progress*. Edison would supply the ideas about inventions, and Lathrop would create the story. Edison prepared many pages of notes for Lathrop, but the project fell apart because he could not find sufficient time away from his inventive work. Lathrop eventually channeled the notes he already had into "In the Deep," a short story serialized by American newspapers in 1896 and 1897.

While Edison proved to be a reluctant collaborator on fictional accounts of the future, he was willing over the years to offer his own speculations in more serious newspaper and magazine interviews. He touched on such subjects as the future of cities, medicine, alternative energy, and machines and labor.

SCIENTIFIC AMERICAN. Edison had a long and mutually beneficial relationship with *Scientific American*, one of the premier English-language journals of the late nineteenth century. The journal started in 1845 as a weekly focused on new inventions (its subtitle was *The Advocate of Industry and Journal of Mechanical and Other Improvements*). It first noted Edison in its 5 September 1874 issue as "well known as a telegraph engineer of the highest ability, and the inventor of a larger number of electrical devices, probably, than any other person living" (*Scientific Amer-*

ican 31 [5 September 1874], 145). **Edward Johnson** announced the **phonograph** in the pages of the 17 November 1877 issue, and several weeks later Edison, Johnson, and **Charles Batchelor** performed the first public demonstration of the device at *Scientific American*'s **New York City** office. The journal's explanatory articles and illustrations helped publicize other Edison creations, including his **tasimeter**, microphone, **electromotograph telephone receiver**, ore (iron) **milling** and **ore (gold) refining** machines, and **electric light and power system**. *Scientific American* sometimes provided more information than Edison wished, as when it published drawings and instructions for making a phonograph. It also served as a forum for his high-profile disputes with **Henry Morton** and **Edward Weston** about the efficiency of his electric lamp and dynamo. *Scientific American* remains (as of 2022) the oldest continuously published magazine in the United States.

SERRELL, LEMUEL WRIGHT (1829–1899). Edison's first patent attorney, Serrell learned the trade from his father, a civil and mechanical engineer who also handled patent cases. The younger Serrell represented several leading telegraph inventors and took Edison as a client after **Marshall Lefferts** introduced them in 1870. He continued as Edison's attorney for a decade as the young inventor established his reputation. In January 1880, Edison put **Zenas Wilber** and George Dyer in charge of his new applications due in part to dissatisfaction with the handling of his increasingly complex foreign specifications. However, he continued to rely on Serrell to act for him with respect to prior applications and interferences.

SHAKESPEARE, WILLIAM (1564–1616). The English playwright was popular in American theater in the mid- and late nineteenth century, and Edison attended performances in Cincinnati while he worked there. Among the productions was *Richard III* in 1867. A telegraph journal (*The Operator*) later stated that the play's title character was "said to have been [Edison's] favorite character, and whenever his duties in the office permitted, he would arise from his instrument, hump his back, bow his legs and proceed with 'Now is the winter of our discontent,' to the great amusement of his fellow-operators" (Israel 1998, 29). Edison would later doodle this famous line in his laboratory notebooks. He considered Shakespeare an important source of inspiration for his own creativity, later saying of the bard, "Ah Shakespeare! That's where you get the ideas! My, but that man did have ideas! He would have been an inventor, a wonderful inventor, if he had turned his mind to it" (Israel 1998, 29). *See also* THEATER.

SIEMENS, WERNER VON (1816–1892). A preeminent electrical inventor, engineer, and entrepreneur, Siemens was the patriarch of the Siemens family and its constellation of manufacturing enterprises in Germany and Great Britain. He used his prestige to advocate for government support of research in the natural sciences and industrial technology. Born into modest circumstances as Ernst Werner Siemens (the honorary particle "von" came in 1888), Siemens studied mathematics and physical sciences in the Prussian military, then turned to invention when his service ended. In 1847 (Edison's birth year), he cofounded Siemens & Halske, a Berlin firm that grew from manufacturing submarine telegraph cables into a giant producer of telephones, railway signals, and electric lighting components, such as dynamos, meters, cables, and arc and incandescent lamps. Siemens's pioneering work in **telegraphy** and electricity—he discovered the dynamo principle for self-exciting **generators**—influenced Edison, who adapted Siemens relays and armature windings into his own telegraph and lighting systems. The Edison and Siemens firms competed in some areas (notably incandescent lamps) and cooperated in others (dynamos and cables), and the two men got along well. When Edison visited Germany during the **Paris Universal Exposition (1889)**, Siemens hosted him and his traveling party, and Edison later provided the German government with a small number of phonographs through Siemens. Decades after Siemens died, the international science community honored him by adopting the *siemens* as the derived unit of electrical conductance.

SILVER LAKE CHEMICAL WORKS. Edison built this manufacturing complex in an area known as Silver Lake, a few miles east of his **laboratory** in **West Orange, New Jersey**. He began acquiring property there in late 1888 and started building two factory buildings the next spring. One made his new **battery (primary)**, the other wax for his phonograph cylinders; together, they marked the beginning of the **Edison Manufacturing Company's** production facilities. The complex expanded over time, with additions including a 1905 factory for chemicals used in Edison's new **battery (storage)**. When World War I interrupted imports from Europe, Edison further enlarged the works to produce vital chemicals for his batteries and phonograph cylinders. The Silver Lake complex also included a research department that tested materials and products made there. The site was redeveloped for unrelated commercial use in the latter half of the twentieth century.

SLOANE, MADELEINE (EDISON) (1888–1979). Madeleine was the first child born to Edison and **Mina Miller Edison**. Her arrival at the end of May 1888 coincided with Edison's push to "perfect" the **phonograph**—the invention he called his "baby"—in the **laboratory** in **West Orange, New Jersey**, and newspapers reported that he used his mechanical "baby" to record the cries of the real one. Madeleine attended the Oak Place School in Akron, Ohio, and then Bryn Mawr College for two years without graduating. She married John Eyre Sloane in 1914 over her parents' objections to his Catholic faith; their four sons were Edison's only grandchildren. In 1938, Madeleine made a brief run as a reformist Republican for a seat in the U.S. House of Representatives. During World War II, she was involved in blood collection by the New Jersey Red Cross. After her mother died, Madeleine took over administration of the Edison birthplace in **Milan, Ohio**.

SPRAGUE, FRANK JULIAN (1857–1934). An electrical engineer, Sprague made significant contributions to Edison's **electric light and power system** in 1883–1884 before developing his own electric motors and a pioneering railway system. Sprague was a naval ensign and recent graduate of the U.S. Naval Academy when he secured an assignment to the **London Crystal Palace Exhibition (1882)**. There he met **Edward Johnson**, who persuaded Edison to hire him in the spring of 1883. Edison assigned Sprague to work on **village plant electrical systems**, which would soon be the job of the **Edison Construction Department**. After helping start the **Sunbury, Pennsylvania**, station, Sprague created mathematical tools to model the conductors for future central stations, obviating the laborious task of building miniature physical layouts. He also helped open other Edison plants and diagnosed faults in the Edison voltage regulator, an essential central station instrument.

Yet Sprague was most interested in **electric motors**, a relative backwater in terms of engineering development. When Edison asked him in April 1884 to take up the subject on his behalf, Sprague resigned to have the freedom to work out his own ideas. Later that year, he formed the Sprague Electric Railway and Motor Company (SERM) with Johnson as president, enabling the company to contract its manufacturing to the **Edison Machine Works**. Sprague created a highly original motor that he used to electrify an urban railroad in Richmond, Virginia, in 1887. The success of the Richmond line created both a market for motors and a rivalry with Edison. Edison disparaged the motor as derivative of his own and came to see Sprague—reputedly an abrasive person—as "a galling thorn in the side of all the boys" (*TAEB* 9:329). It did not help that Sprague had (at Johnson's solicitation) produced an 1886 report recommending that the **Edison Electric Light Company** develop an **alternating current (AC)** distribution system. Sprague resented the takeover of his company by **Edison General Electric (EGE)** in 1889–1890 and the subsequent rebranding of its products as "Edison" (EGE later marketed different motors developed by Edison). He severed all connections with SERM and EGE. Sprague enjoyed additional success with streetcars, hoists, and

elevators but did not attain his goal of electrifying U.S. mainline railroads.

STIERINGER, LUTHER (1846–1903). Formerly a gas engineer, Stieringer joined Edison's **laboratory** in **Menlo Park, New Jersey**, in 1881—the start of a distinguished career as an illuminating engineer. He adapted gas fixtures for incandescent **electric lamps** and was responsible for the artistic lighting of several Edison exhibitions, including at the Southern States exhibition in Louisville, Kentucky, in 1883, which was the largest electric light display to that time. Later, as a contractor for Edison, he had a major role in mounting large displays at the **Philadelphia International Electrical Exhibition (1884)**, in **New York City** and Minneapolis (1890), and at the **World's Columbian Exposition (1893)** in Chicago. Stieringer at one time was half of **Bergmann & Company**'s Wiring Department. He supervised the installation of lighting fixtures at **Glenmont** (Edison's home), contributed to the design of Edison's municipal lighting system, and took out several patents in his own name.

STOCK TICKER. The stock ticker, a device for printing stock prices on a continuous narrow strip of paper, was once synonymous with **New York City**'s financial district. Credit for this iconic machine goes not to Edison but to Edward Calahan, who devised the first small printing telegraph for stock prices in 1867. Calahan's printer became the foundation of the **Gold and Stock Telegraph Company**, which enlarged its reporting services to include the prices of commodities such as gold, cotton, oil, and produce. Edison began working on printing telegraph technology in **Boston** in 1868. Following his move to New York the next year, he developed a printer for the Gold and Stock Reporting Company (a company soon acquired by Gold and Stock Telegraph). After designing another printer with **Franklin Pope** for their Financial and Commercial Telegraph Company, Edison began working as a contract inventor for Gold and Stock. He created his Universal stock printer in 1871, not long before **Western Union** acquired Gold and Stock. His improved ticker made use of several

Universal Stock Printer, 1871. *Courtesy of Thomas Edison National Historical Park.*

key contributions to printing telegraphy, chief among them a screw-thread unison that synchronized all tickers on a telegraph line so that they printed the same information from a single transmitter. Edison also developed a private-line version of the printer that used a keyboard to transmit messages.

STREET LIGHTING. Beginning in the 1870s, electric arc lamps began to replace gas lamps for illuminating thoroughfares and public spaces in many cities. In these systems, an electric current jumped between two carbon rods, creating an intensely bright arc or continuous spark and gradually vaporizing the rods. Municipal arc lighting systems typically operated at many times the voltage of the Edison incandescent systems (depending on the number of lamps), which Edison thought too dangerous. As a safer alternative, he developed a "municipal system" that used small currents (4 amperes, though still at relatively high pressure—1,200 volts) to power low-resistance incandescent lamps wired in series. Each lamp had an automatic switch (the "cutout") to maintain the circuit in case it failed.

Edison patented the system in 1884, and the first installations took place the following year. **Edison General Electric** used the system until 1892, but successor **General Electric Company** adopted the **Thomson-Houston Electric Company** arc light instead.

SUNBURY, PENNSYLVANIA. In 1883, Edison completed his first **three-wire system** electric light central station in this small city, about 55 miles up the Susquehanna River from Harrisburg. After overseeing the later stages of construction and attending its inauguration on 4 July, he stayed in town another eight days to monitor operation and train staff. Built under the aegis of the **Edison Construction Department**, the 500-light plant was a template for Edison's **Village Plant** business and the spread of Edison incandescent lighting beyond the **New York City** metropolis. Its difficulties also presaged those at similar plants. Among the problems were shoddy construction, inept or inattentive operators, and insufficient capital to pay its debts to Edison or finance expansion. The plant operated well into the twentieth century.

SWAN, JOSEPH WILSON (1828–1914). Swan was a practical chemist by trade and a pioneer of photographic technologies whose work in electric lighting made him an early rival of Edison in Great Britain. He started experimenting in the early 1860s, briefly bringing a spiral of carbonized cardstock to a glowing heat in an evacuated bulb. Soon after Edison's announcement of a successful incandescent **electric lamp** in late 1879, Swan claimed to have already made similar lamps. Then, in October 1880, Swan claimed to have used an improved vacuum pump in 1877 to make a lamp with a thin carbon filament—two years before Edison's. Edison and his allies (notably **Edward Johnson**) pointed out that the public record showed nothing of the kind, that lamps Swan demonstrated in 1878 and 1879 had carbon rods and low resistance and were connected in series, and that even in 1880, Swan seemed unaware of the practical necessity of using **high resistance** and connecting lamps in parallel. But Swan enjoyed support from Britain's scientific fraternity, and Edison took care not to antagonize him or the Swan Electric Light Company. Swan was gracious, congratulating Edison on his prize at the **Paris International Electrical Exposition (1881)**. The inventive dispute ceased to matter for legal purposes after October 1883, when the Swan and British Edison firms merged into a new company named for both men. Swan continued to work in electric lighting and made significant contributions, notably processes for creating superior filaments of extruded cellulose. Britons still regard him as at least a coinventor of the incandescent electric lamp.

T

TALKING DOLL (TOY PHONOGRAPH). The talking doll was Edison's unsuccessful attempt to incorporate the **phonograph** into a child's toy. Simulating vocalized sounds—including human speech—by mechanical means such as reeds, bellows, and so on, was an old idea, and a talking toy was among Edison's first suggested uses for the phonograph in 1878. However, he soon threw himself into electric light research, and everything related to the phonograph languished for nearly a decade.

In the summer of 1887, William Jacques, a PhD physicist working for the **American Bell Telephone Company**, approached Edison about licensing phonograph patents for a talking doll he had created. Edison agreed, and Jacques and a business partner formed the **Edison Phonograph Toy Manufacturing Company (EPTMC)**. In early 1888, taking Jacques's device as a starting point, **Charles Batchelor** began to design a practical miniature phonograph. Creating one to withstand a child's rough handling was difficult. Working intermittently, Batchelor tried recording on small metal or wax cylinders and worked on various reproducing mechanisms. The next year passed without a marketable model, though the company did send a prototype to the **Paris Universal Exposition (1889)**. Edison made considerable efforts to secure a supply of high-quality European doll bodies (including sending **Albert Dick** to canvass manufacturers) and to establish sales arrangements for foreign countries. The EPTMC accepted Edison's prototype in September 1889, but it was only the following February, after several false starts,

that the **Edison Phonograph Works** began making tiny crank-operated phonographs and assembling them into twenty-two-inch doll bodies. About eighteen girls and women (some possibly recruited through a New York talent agency) recorded a dozen nursery rhymes one wax cylinder at a time, as there was no way to duplicate recordings. Approximately 2,500 dolls went on the market at the hefty price of $10 to $25 each, but a high percentage were returned with defects. They represented a small fraction of the more than 10,000 units made, most of which simply went into storage at the factory. Manufacture ceased in May 1890. In October of that year, Edison tried to take over the company with a board that would institute his policies, especially regarding foreign markets. When that gambit failed, Edison washed his hands of the doll and tried to recover whatever costs he could from the company, which answered by suing him for breach of contract. (The suit was later dropped amid counterclaims by Edison.) The National Park Service has made eight doll recordings available at https://nps.gov/edis/learn/photosmultimedia/hear-edison-talking-doll-sound-recordings.htm.

TASIMETER. At the suggestion of astronomer Samuel Langley, Edison devised this extraordinarily sensitive instrument to measure tiny fluctuations in temperature. He introduced it to the press in May 1878 while still searching for an appropriate name. Favoring words made from classical roots ("phonograph"), he chose "tasimeter," a combination of the Greek words for "extension" and "measure."

Like the **carbon-button telephone transmitter**, the tasimeter used the variable-resistance property of carbon to regulate an electric circuit. It had a vulcanized-rubber rod that, when subjected to tiny increases in heat, expanded against one of two metal plates. Between the plates rested a carbon button in circuit with a battery and a galvanometer (a visual current-measuring instrument). As pressure on the plate changed, so did the resistance of the carbon—and therefore the current to the galvanometer. For astronomical uses, the instrument was connected to a telescope that focused light from distant bodies onto the rubber rod.

Encouraged by leading astronomers, Edison took the tasimeter on an 1878 solar eclipse expedition to Wyoming for the purpose of measuring the heat of the sun's corona. On the way back to New Jersey, he presented his results at the annual meeting of the **American Association for the Advancement of Science** in St. Louis. He reported measuring the heat from the giant star Arcturus but was less successful with the eclipse. Overcoming difficult conditions, he got the corona in focus only to find that the tasimeter's sensitivity exceeded that of the galvanometer, whose scale was not wide enough. Despite his failure to collect useful solar data, Edison considered the tasimeter a success and hoped it could lead to the discovery of stars too faint for visual observation. To encourage its use, he arranged for its manufacture and sale without royalty to him. A few scientists bought tasimeters, but a consensus developed that it was too difficult to use and insufficiently precise for serious research, and Edison abandoned the instrument. *See also* ASTRONOMY.

TATE, ALFRED ORD (1863–1945). Tate was Edison's private secretary for seven years, managing the inventor's voluminous correspondence and helping to oversee his financial and business affairs. A native of Peterborough, Ontario, Tate had worked as a telegraph operator and train dispatcher when, in May 1883, he was hired (through an acquaintance of **James MacKenzie**) to assist **Samuel Insull** with the flood of office work related to the Edi-

son Construction Department. He later acted as an agent in Canada for the Edison lighting interests, particularly the **Edison Machine Works**. Drawing on his Canadian contacts and telegraphic experience, Tate set up the first commercial demonstration of the **phonoplex telegraph** and managed the considerable business that resulted. Edison deemed him "a rising man [who] is honest smart & ambitious" enough to run the **Edison Lamp Company** but instead chose him to succeed Insull as private secretary in 1887 (*TAEB* 8:685). After moving into the new **laboratory** in **West Orange, New Jersey**, Tate supervised the office staff and worked with Insull to manage the tight flow of cash among the laboratory and Edison's various enterprises. He took charge of promoting and selling **batteries (primary)**, laying the foundation for a business that would be important to Edison for decades. Tate was also central to the unsuccessful effort to market the **talking doll**, served as an executive of the **North American Phonograph Company**, and arranged Edison's historical exhibit at the **World's Columbian Exposition (1893)**. He left Edison's direct employ in 1894 to work in battery development and manufacturing. Forty-four years later, he published *Edison's Open Door*, a colorful memoir and selective biography of Edison.

TELEPHONE. *See* CARBON-BUTTON TELEPHONE TRANSMITTER; ELECTROMOTOGRAPH TELEPHONE RECEIVER; MICROPHONE CONTROVERSY.

TESLA, NIKOLA (1856–1943). A pioneering electrical inventor and engineer, Tesla made major contributions to the development of **alternating current (AC)** power systems, for which he is sometimes misremembered as an opponent of Edison. Tesla was born to Serbian parents in present-day Croatia but spent most of his working life in the United States. Before emigrating, he studied mathematics and physics and gained technical experience at a Budapest electrical firm and at Edison lighting companies in and around Paris. Tesla came to **New York City** in 1884 and worked about six months at the **Edison Machine Works**, mainly on **generator** design. Legend has it that he left after being denied a bonus promised by Edison (who barely met him) or **Charles Batchelor**, but that story is incorrect. (In his autobiography, Tesla blamed the Machine Works superintendent for his disappointment.) According to his most recent biographer, Tesla probably left after the company declined to buy an arc light he had developed. Tesla then set up a **laboratory** in New York, where he designed and patented arc lights, generators, transformers, and highly original polyphase motors. In 1888, he sold his basic AC patents to **George Westinghouse** and joined Westinghouse's engineering team in Pittsburgh, Pennsylvania. Tesla spent only about a year there—and played no direct role in the so-called **War of the Currents**—but the motor designs he and his colleagues developed in the late 1880s revolutionized the electrical industry by making it possible to use AC for electric power instead of lighting alone. **Westinghouse Electric Company** engineers developed Tesla's polyphase induction motor and transmission system, which the company used to light the **World's Columbian Exposition (1893)** in Chicago and then to distribute power from Niagara Falls in 1895.

Returning to work on his own, Tesla made a series of intriguing inventions and discoveries. Investigating the wireless transmission of electric power, he partially anticipated the work of Guglielmo Marconi in radio. He made the so-called Tesla coil, which created spectacular visual effects by transforming a low-voltage, high-frequency input into an extremely high-voltage output. Edison and Tesla maintained cordial relations, belying later attempts to recast them as bitter rivals. When fire destroyed Tesla's New York laboratory in 1895, Edison offered him the use of his own facilities. And when Tesla received the Edison Medal from the **American Institute of Electrical Engineers** in 1917, he reflected on having worked for "this wonderful man" as an immigrant to New York (Morris 2019, 180 n.).

THEATER. Edison had a lifelong fondness for the theater, one of his few diversions from work. As a young telegraph operator, he attended performances whenever he could. He developed a taste for Shakespeare and would later see the playwright's inventiveness as an inspiration for his own. Among his favorite plays were *Othello* and *Richard III*, whose famous opening lines ("Now is the winter of our discontent . . .") he used in early phonograph demonstrations and sometimes doodled on notebook pages. Edison continued going to the theater (and seeking out Gilbert and Sullivan light operas) despite his growing deafness.

Theaters were early and important customers for incandescent electric lighting. Compared with gas, incandescent lights were safer, quieter, and more comfortable. They also allowed colored illumination and permitted better control of their effects. Edison pushed the **Edison Lamp Company** to develop special bulbs for the stage, but it was rival **Joseph Swan** who in 1881 provided bulbs for the first fully electric theater in the world—London's Savoy, best known for its staging of Gilbert and Sullivan operettas. Edison personally assisted the **Edison Company for Isolated Lighting** when it installed a system in 1882 at **Boston**'s Bijou Theater, the first American theater lit by electric incandescence. He and **Luther Stieringer** designed lighting for the new Lyceum Theater in **New York City** in 1884. Edison took charge of the installation of an isolated electric plant at Niblo's Garden theater in New York in preparation for the 1883 production of *Excelsior*, an allegorical ballet of "progress" in which dancers carried wands tipped by small incandescent lamps.

Edison personally joined a theatrical production of another sort in 1884: an outdoor political spectacle for presidential candidate James Blaine. He rode at the head of a Blaine procession down New York's Fifth Avenue, leading 350 uniformed men carrying sixteen-candlepower electric torchlights and wearing hats with smaller lights on the plumes. (Blaine lost anyway.)

THEOSOPHICAL SOCIETY. In 1875, a **New York City** resident and Ukrainian-born noblewoman named Madame Helena Petrovna Blavatsky cofounded the Theosophical Society. Theosophy drew on the occult, ancient scriptures, Hinduism, Buddhism, and writings of European philosophers to posit that all men were immortal brothers, living and reliving their lives through reincarnation. Society cofounder and president Henry S. Olcott wrote Edison in March 1878 in hopes of introducing him to Blavatsky, who "would not like to miss the pleasure of seeing so thoroughbred a Heathen—as you say you are" (*TAEB* 4:176; Blavatsky had mentioned Edison's 1875–1876 **etheric force** experiments in her 1877 book *Isis Unveiled*). Olcott visited Edison in the spring of 1878, and Edison joined the Society in April. Edison never met Blavatsky, and his participation in theosophy was limited to waiving the royalty on a phonograph for the Society and providing copies of his portrait to its branches. In December 1878, Olcott and Blavatsky formed a branch of the Society in London; they later established the Society's world headquarters in India. After Blavatsky's death in 1891, the Society splintered into factions. The American Theosophical Society headquarters in Wheaton, Illinois, is named for Olcott.

THOMAS A. EDISON INCORPORATED (TAE INC.).

This conglomerate was created in 1911 to bring most of Edison's disparate business enterprises under a single administrative structure. It was formed by changing the corporate title of the **National Phonograph Company** and by acquiring the Edison Business Phonograph Company and the **Edison Manufacturing Company**. All the remaining Edison companies (except for the **Edison Portland Cement Company**) were later brought into TAE Inc., including the **Edison Phonograph Works** (1924), the **Edison Storage Battery Company** (1932), and the **Emark Battery Company** (1933). Edison served as president until 1926 (when his son **Charles Edison** took over day-to-day management) and remained as chairman of the board until his death in 1931. TAE Inc. was structured as a multidivisional company with divisional managers for each product line. These managers reported to Edison's leadership team, who formed a central administrative body called the Executive Committee. TAE Inc. merged with the McGraw Electric Company in 1957 to form the McGraw-Edison Company, which was itself acquired by Cooper Industries in 1985. *See also* MAMBERT, STEPHEN.

THOMAS A. EDISON PAPERS. In the mid-1970s, leaders of cultural and educational institutions made plans to celebrate the 1979 centennial of Thomas Edison's incandescent **electric lamp**. The initiative was led by John T. Cunningham, chairman of the New Jersey Historical Commission, and Arthur Reed Abel, archivist at the Edison National Historic Site (now the **Thomas Edison National Historical Park [TENHP]**). An inter-institutional committee was formed that included representatives of those organizations and the Smithsonian Institution. With financial support from the Edison Electric Institute, the group commissioned a report from James Brittain, a distinguished historian of electrical technology. Brittain recommended establishing an Edison Papers project with the goal of publishing a selective book edition and supplementary microform. He also recommended that the committee identify a major academic home for the project. That honor went to Rutgers University.

Rutgers conducted a nationwide search for a new faculty member to serve as the project's editor in chief. In 1978, Reese Jenkins, a historian of technology and science at Case Western Reserve University, was selected. Representatives of Rutgers, the New Jersey Historical Commission, the National Park Service, and the Smithsonian signed a cosponsorship agreement that same year. (Those institutions, among others, continue to provide oversight.) The project established editorial offices at the Rutgers flagship campus in New Brunswick, New Jersey, and at the present-day TENHP in **West Orange, New Jersey**. It hired historical editors and researchers who, under the leadership of Professor Jenkins, developed plans for two complementary series of publications. One was a selective microform edition consisting of images of a modest percentage of the available documents, chosen according to their intrinsic historical interest and ability to represent the body of archival materials for serious researchers. The other was an even more selective hardcover print edition of transcribed and fully annotated documents intended for a broad range of scholars and others interested in Edison. The project also aimed to create ancillary materials and activities to expand the general public's understanding of Edison's work and life.

Robert Rosenberg followed Jenkins as director in 1995 and launched the project's digital image edition five years later. He was succeeded in 2002 by Paul Israel, a historian of technology and Edison scholar who still leads the project and has overseen a major expansion and update of the digital edition.

The scale of the Edison Papers is enormous. In his feasibility study, James Brittain estimated that the TENHP archive contained approximately 1.5 million pages, with other repositories possibly holding several thousand additional pages of documents. Ten years later, an inventory of the West Orange collections tallied more than 5 million pages. The project has since collected nearly 20,000 documents from other repositories. Consequently, what was conceived as a twenty-year editing project has grown into a much more complex undertaking now scheduled for completion fifty-plus years after it began.

The Edison Papers digital edition (https://edisondigital.rutgers.edu) contains scanned images of more than 153,000 documents (such as correspondence, laboratory notebooks, financial records, and legal materials) accessible through a variety of search tools. The documents are drawn largely from the five-part microfilm edition but include more than 15,500 gathered from more than 100 archives and private collections. Selected archival materials from the period 1920–1931 will be added in the future. The book (print) edition, now available in open-access formats on Project Muse (the digital platform of Johns Hopkins University Press), consists of nine volumes of 3,458 transcribed and fully annotated documents covering 1847–1889. Six more volumes are planned to cover the remainder of Edison's life.

Major funding for the Edison Papers project comes from Rutgers University, the National Endowment for the Humanities, the National Historical Publications and Records Commission, the New Jersey Historical Commission, and private grants and donations.

THOMAS EDISON BIRTHPLACE MUSEUM. Edison's father **Samuel Edison** built this one-and-a-half-story brick house in Milan, Ohio, in 1841. Edison lived here from his birth in 1847 until the family moved to **Port Huron, Michigan**, in 1854. Edison's sister (**Marion Edison Page**) repurchased the property in 1894, and he inherited it a dozen years later. After Edison died, **Mina Edison** and their daughter **Madeleine Edison Sloane** transformed the birthplace into a museum, which opened to the public in 1947. The house was restored as nearly as possible to its original appearance, furnished with family mementos and period household articles. Now on the National Register of Historic Places, the house is maintained by the Edison Birthplace Association Incorporated, a private, nonprofit organization.

THOMAS EDISON CENTER AT MENLO PARK. The first effort to commemorate the site of Edison's **laboratory** complex in **Menlo Park, New Jersey**, occurred in 1925, when the **Edison Pioneers** dedicated a bronze tablet on the overgrown grounds near the present-day Metropark train station. The group had initially hoped to raise a seventy-foot monument but lacked funds; however, the idea inspired later plans. In advance of Light's Golden Jubilee—the fiftieth anniversary of Edison's incandescent **electric lamp**—a 120-foot steel tower was erected on the site of the main laboratory building. It had an incandescent "eternal light" at its base and a large incandescent light at the top, first lit on 21 October 1929. The Pioneers and the Edison family worked with the state of New Jersey to acquire twenty-four acres around the site to create Edison State Park, dedicated in 1937. That same year, a Wilmington, Delaware, architectural firm (Massena & duPont) began designing a permanent light tower to replace the steel one. The resulting art deco structure, 131 feet high, is clad in precast concrete panels created by John Joseph Earley, a pioneer in the use of architectural concrete. The panels are made of Edison Portland cement, light-reflecting quartz, and ceramic aggregate. At the top sits a four-teen-foot electric lightbulb built by the Corning Glass Works, which supplied glass for Edison's lamps. The "eternal light" is in a room at the bottom. The Edison Memorial Tower was dedicated on Edison's birthday (11 February) in 1938 and rededicated in 2015 after extensive renovations. A two-room museum devoted to artifacts and interpretive exhibits about the laboratory now occupies the original gatehouse. The Thomas Edison Center at Menlo Park and the surrounding park are jointly administered by the New Jersey Department of Environmental Protection's Division of Parks and Forestry, the Township of Edison, and the nonprofit Edison Memorial Tower Corporation.

THOMAS EDISON HOUSE. Edison lived in the historic Butchertown neighborhood in Louisville, Kentucky, while working as a telegraph operator in 1866–1867. He likely rented a room in this 1850 shotgun duplex on East Washington Street or in a similar building nearby. The house features a small museum about Edison and his inventions.

THOMAS EDISON NATIONAL HISTORICAL PARK. This national park includes two properties: Edison's **laboratory** complex **in West Orange, New Jersey**, and **Glenmont**, his home in nearby **Llewellyn Park**. The National Park Service acquired the twenty-one-acre site and its contents through a series of agreements with **Thomas A. Edison Inc.** (later the McGraw Edison Company) between 1955 and 1962. Designated at first as the Edison Laboratory National Monument, it was renamed the Edison National Historic Site in 1962. It received its current designation in 2009. Besides the buildings, the complex houses thousands of Edison artifacts and photographs and an archive with an estimated 5 million pages of documents.

THOMSON, ELIHU (1853–1937). *See* THOMSON-HOUSTON ELECTRIC COMPANY; UNKNOWN NATURAL FORCES.

THOMSON, SIR WILLIAM (LORD KELVIN) (1824–1907). A Cambridge-educated professor at the University of Glasgow, Thomson embodied the British scientific establishment whose respect Edison sought. Preeminent as both a telegraph engineer and an academic mathematician and physicist, Thomson played a key role in laying and opening the transatlantic cable (for which he was knighted in 1866). His practical inventions included a mirror galvanometer equally useful in a telegraph office or a physics laboratory. Thomson's scientific interests ranged across electricity and energy, but his appreciation of technology shaped his style of research. Like **Michael Faraday**, he believed in grounding knowledge in experience gained from experiments or models—an approach consistent with Edison's own empirical bent. It was high praise, then, when a prize committee headed by Thomson at the 1876 **Centennial Exhibition** cited the **electric pen** for "exquisite ingenuity" (Israel 1998, 125) and when Thomson singled out Edison two years later as "the very first Electrician of the Age" (*TAEB* 4:140). Thomson visited Edison's new **laboratory** in **Menlo Park, New Jersey**, after the 1876 exhibition.

Edison turned to Thomson for support in the 1878 **microphone controversy** about inventive priority for that device. However, Thomson saw the matter as a case of simultaneous invention and gave neither Edison nor David Hughes credit for the underlying principle. He did fault Edison for making a "violent attack in the public Journals" instead of letting the technical community resolve the matter quietly (*TAEB* 4:428). Yet, unlike some of his British colleagues, Thomson was not prejudiced by Edison's bluster, and he kept an open mind about the feasibility of incandescent lighting in general and Edison's lamp in particular. In 1881, he asked Edison to provide lamps for a private comparison with those of British rival **Joseph Swan**. (He found them to be similar.) Thomson was elevated to the peerage as Baron Kelvin in 1892, the first scientist so honored. He visited Edison again at **West Orange, New Jersey**, a decade later.

THOMSON-HOUSTON ELECTRIC COMPANY. This firm was one of the three major electrical manufacturers in the United States by the mid-1880s, along with the **Westinghouse Electric Company** and the **Edison Electric Light Company**. It was created in 1883 to make and sell arc lighting equipment patented by Elihu Thomson and **Edwin Houston**, Thomson's former mentor and colleague at Philadelphia's Central High School. (Thomson had recently severed his affiliation with an earlier company created for the same purpose.) It had offices in **Boston** and a factory complex in nearby Lynn, Massachusetts. The company dominated the arc lighting market on the technical merits of its equipment (especially Thomson's self-regulating generator) and the sales and administrative acumen of its head, Charles Coffin, a former shoe manufacturer. In the mid-1880s, it entered the incandescent and isolated lighting fields by acquiring or licensing patents, including those of **William Sawyer**. It also began to make and market motors (including those for streetcars) and **alternating current (AC)** equipment and to build central stations, becoming a formidable competitor in each field.

Thomson-Houston prospered from patent-licensing agreements with Westinghouse and in early 1889 sought an even stronger alliance with the **Edison General Electric Company (EGE)**. It essentially offered to divide up the lighting business and grant EGE the use of its AC system. EGE president **Henry Villard**, who had made his own overtures to Thomson-Houston, appeared willing, but Edison refused. Thomson-Houston overcame intense competition to become more profitable than EGE due largely to its superior organization, sales force, and financing. The growing imbalance and the promise of savings through consolidation caught the eye of each company's bankers. In early 1892, **J. P. Morgan** and his Boston counterparts arranged a merger of the two firms into a new concern, the **General Electric Company**—one of the nation's largest companies. Having lost their leverage, Villard and Edison could accept the consolidation only as a fait-accompli. Personal names disappeared from the company's identity, Villard resigned to make way for Coffin as president, and **Samuel Insull**—formerly EGE's de facto manager—left for a new career in Chicago.

THREE-WIRE SYSTEM. Edison modified his two-wire system of electrical distribution to reduce the cost of building central station service districts. The result was a three-wire system (created independently of a similar plan by British electrical engineer **John Hopkinson**) in which every lamp could still operate on its own. Copper conductors were among the costliest parts of central station systems, and their expense in the pioneering **Pearl Street Station ("First District")** dampened investors' enthusiasm for other big plants. The three-wire plan reduced copper use by nearly two-thirds. It allowed current to flow from the dynamos at double the 110 volts used by lamps and motors, meaning that a smaller current—requiring smaller conductors—could provide the same amount of electrical energy. A bit of electrical sleight of hand allowed 110-volt lamps and motors to operate safely from this 220-volt supply without recourse to transformers or other voltage-reduction devices. Each main circuit was divided into two parallel sub-circuits by a third wire between the main conductors, giving half the voltage of the whole to each sub-circuit (where the lamps and motors were). Because the third (or "compensating") line could balance the two halves without carrying much current itself, it could be quite thin. The three-wire system became standard for Edison central stations built after mid-1883 (including those put up by the **Edison Construction Department**), and the original Pearl Street district was retrofitted starting in 1890. Edison explored four- and five-wire systems as alternatives to the more economical distribution of **alternating current**, but these were not practical. *See also* SUNBURY, PENNSYLVANIA; VILLAGE PLANT SYSTEM.

THURSTON, ROBERT HENRY (1839–1903). As director of Cornell University's Sibley College of Mechanical Engineering and the Mechanic Arts from 1885 until his death, Thurston transformed the school into an academic center for both mechanical and electrical engineering. (John Vincent Miller, brother of **Mina Miller Edison**, studied there in the late 1890s.) Renowned as an expert on steam engines, Thurston sought to upgrade Sibley's electrical resources on his arrival in Ithaca from his prior post at the Stevens Institute of Technology. He persuaded Edison to donate a dynamo, incandescent lamps, and other equipment. Although Edison resisted Thurston's 1889 request for a large monetary donation for a new electrical laboratory, Thurston later acknowledged his unspecified contributions to the laboratory's mission. Thurston also served on the jury of the **Paris Universal Exposition (1889)** that named Edison a recipient of its highest honor.

TOMLINSON, JOHN CANFIELD (1856–1927). A partner in the **New York City** law firm Ecclesine & Tomlinson, Tomlinson served as Edison's business and personal attorney from 1883 to September 1888. He was also a patent attorney for the **Edison Electric Light Company.** Tomlinson earned bachelor's (1875), bachelor of law (1877), and master's degrees from the University of the City of New York. Entering the Edison world through his firm's investigation of contracts for lighting the Brooklyn Bridge, he quickly earned the inventor's respect and friendship. But their bond was shattered after Tomlinson negotiated the sale of Edison's phonograph rights to **Jesse Lippincott** in June 1888. Tomlinson and **Ezra Gilliland** urged a reluctant Edison to take the deal, one from which they stood to profit individually by receiving large amounts of stock in Lippincott's prospective **North American Phonograph Company.** Edison claimed that he learned of the lucrative side agreements only when Lippincott could not make a scheduled payment to him a few months later. He then became enraged by what he considered an underhanded scheme for Gilliland and Tomlinson to enrich themselves at his expense. Edison promptly fired Tomlinson and had Edison Electric dismiss him. He also filed a lawsuit against both men (who left the United States for several months); the suit dragged on for several years but did not produce any settlement. Tomlinson was subsequently a partner in the Manhattan law firm Tomlinson, Herrick, Hoppin & Coats until his death.

TOPSY. The electrocution of Topsy, a circus elephant at Luna Park on Coney Island, has been erroneously linked to Edison. According to popular mythology, Edison had her killed by **alternating current (AC)** as a publicity stunt in the **War of the Currents** to demonstrate the inherent dangers of AC. Although Topsy was deliberately electrocuted at Luna Park on 4 January 1903, Edison—who had essentially left the electric power business a decade earlier—had no part in the decision or the preparations, and he was not present when it occurred. Abused by her trainer and others, Topsy was condemned after separately killing three men, one of whom had placed a lighted cigarette in her mouth. Luna Park officials decided to euthanize her by hanging, but the local Society for the Prevention of Cruelty to Animals (SPCA) objected to that method. They then proposed electrocution, in combination with poisoning and strangulation, as more humane, and the SPCA agreed. The Edison Electric Illuminating Company of Brooklyn—the local utility in which Edison had no influence—provided the technical means and 6,600 volts of AC power. Newspapers fanned public interest in the proceedings, and spectators gathered at Luna Park to witness it. Topsy was dosed with poison, and a rope was placed around her neck. Once the electricity was switched on, the elephant toppled over and died within ten seconds, obviating the need for the rope.

The myth of Edison's involvement in Topsy's killing probably arose from the fact that the **Edison Manufacturing Company** (of which Edison was president) filmed the event and distributed the film under his name as *Electrocuting an Elephant.* However, Edison had no role in producing "Edison" films at this time, and no evidence has come to light that he had anything to do with this one. **William Gilmore** managed the Manufacturing Company, and his extant correspondence with Edison contains no mention of Topsy. The electrocution also had no bearing on the War of the Currents, which had ended more than a decade earlier in favor of AC, which even the Brooklyn utility company was using by 1903.

TRASK, SPENCER (1844–1909). Trask, an investment banker and philanthropist, provided financing and executive leadership for Edison's electric light and power businesses over the course of two decades. He was a director of the **Edison Electric Light Company** and invested in the **Edison Electric Illuminating Company of New York** in 1881, twice serving as that firm's president (1884–1989 and 1891–1899). He was also an organizer, director, and president of the Edison Electric Illuminating Company of Brooklyn. Trask's investment firm, Spencer Trask & Company, invested in and provided professional services to the **North American Phonograph Company**. A Brooklyn native, Trask attended public schools and the Polytechnic Institute of Brooklyn; he graduated from Princeton College in 1866. Three years later in **New York City**, he established his own investment banking firm, which became Spencer Trask & Company. It specialized in railroad and municipal bonds and acted as the financial agent for Queen Victoria in the United States.

TYPEWRITER. The typewriter was an unstandardized machine with a complex inventive history in 1870, when Edison began adapting it to print messages received by the **automatic telegraph system**. He filed a patent on his improvements in November 1871, but the **Automatic Telegraph Company** instead adopted a typewriter invented by Christopher Sholes. Edison continued to experiment, patenting an electric model in 1872 that was a modification of his Universal private-line printing telegraph. Six years later, he applied for a patent on a design that produced stencils like his **electric pen**. He experimented with stencil typewriters periodically in the 1880s, including for **Albert Dick**, without patenting further improvements. Concurrently, Edison's office staff used typewriters to prepare a growing portion of his own voluminous outgoing correspondence. *See also* MIMEOGRAPH.

U

UNGER, WILLIAM C. (1850–1879). Listed in the 1870 federal census as a telegrapher, Unger became Edison's partner that year in the manufacturing business called the Newark Telegraph Works (later renamed Edison & Unger). His brother George Unger worked as a machinist for the firm. After the partnership dissolved in July 1872, William moved to **New York City** and made telegraph instruments, light machinery, and models. He formed a business with Hamilton Towle the next year. Unger appears to have returned to **Newark, New Jersey**, in 1878 and may have joined his brothers Herman, Eugene, and Frederick (and likely George) in the noted jewelry firm H. Unger & Company (later Unger Bros.). William, George, and Frederick all died of tuberculosis in 1879.

UNKNOWN NATURAL FORCES. The first evidence of Edison's abiding interest in unknown physical forces occurred in April 1874, when he noticed the **electromotograph** phenomenon, which he initially described as a "new force" (*TAEB* 2:178). Edison soon decided that this effect, while interesting and potentially useful, was no such thing. However, his search for such a novelty began in earnest the following year, when he conducted experiments in early June 1875 to discover a new force he could use in telegraphy. It was while experimenting with **acoustic telegraphy** in November 1875 that Edison discovered what he believed to be a new force. Finding that he and his assistants could produce sparks at a distance from an electrical charge, Edison decided that they

had in fact observed a new force, which he soon began calling the **etheric force**—referring to the invisible "ether" supposed to permeate space. After Edison announced his discovery, most electrical experimenters (notably **Elihu Thomson**) decided it was nothing more than an example of induction, a familiar product of **electromagnetism**. Although **James Clerk Maxwell** had developed a mathematical theory of the electromagnetic field that embraced the transmission of high-frequency waves through space, no one at the time recognized that Edison had generated and observed such waves. It was not until Heinrich Hertz experimentally confirmed Maxwell's theory in 1889 that researchers understood the significance of Edison's experiments or saw the possibility of using such waves for radiotelegraphy.

From the beginning of his etheric force experiments, Edison sought to develop an original theory to explain the phenomenon. He argued that all known forms of energy (electricity, magnetism, heat, and light) were interchangeable and could be transformed to (or from) unknown other forms of energy. In 1885, he began a new research program to explore such transformations in the hope of discovering a "new mode of motion or energy," which he called "XYZ" (*TAEB* 8:280, 132). Even after Hertz's decisive demonstrations of radio waves, Edison continued to experiment in the hopes of discovering this heretofore unknown force.

UPTON, FRANCIS ROBBINS (1852–1921). Upton was a mathematician and physicist who received the first master of science degree awarded by the College of New Jersey (now Princeton University). He joined Edison's staff at the **laboratory** in **Menlo Park, New Jersey,** in November 1878 to review patents and technical literature for prior work on electric lighting. Upton's number-crunching abilities and grasp of physical theory proved crucial for the design of **electric lamps** and **electric generators** and especially the requirements of the **electric lighting system** as a whole. His contributions were such that Edison awarded him 5 percent of future electric light profits and royalties and assigned him to oversee the installation of a lighting system on the steamship *Columbia.* Upton was an original member of Edison's lamp factory partnership and, from 1881, general manager of the factory. Initial struggles with cash flow and occasional product quality lapses prompted Edison to scold Upton for "losing the art" and "degenerating into a mere business man" (*TAEB* 8:309). He drove down costs, partly through economies of scale and partly through hard dealing with labor, such as relocating the **Edison Lamp Company** in search of cheaper female workers. Upton had studied in Germany, and Edison entrusted him with some sensitive European business, including one trip that resulted in the **Edison Electric Light Company** acquiring American patent rights on key parts of an **alternating current** distribution system. Upton stayed with the lamp factory until 1895 (following its absorption by the **Edison General Electric Company** and its successor **General Electric Company**). He returned to the fold as an agent for the **Edison Portland Cement Company,** retiring in 1911. In later years, Upton was the first president of the **Edison Pioneers.**

USS *JEANNETTE.* The steamship *Jeannette* was the vessel for a highly publicized 1879 Arctic expedition sponsored by *New York Herald* publisher and editor James Gordon Bennett Jr. At the request of its commander, U.S. Navy Lieutenant George De Long, for some type of electric lighting to break up the long northern nights, Edison supplied one of his early **electric generators,** one or more arc lights, and several platinum-spiral incandescent lamps (predecessors of the carbon-filament **electric lamp**). Edison also helped obtain a steam engine, telephones, and wire for the *Jeannette* before it sailed from San Francisco on 8 July 1879. That fall, after the ship was locked in ice, the crew tried the electric lights but could get no current from the generator. Suspecting that the machine had gotten wet, they tried in vain to reinsulate it before using its parts in makeshift pumps. The ship was crushed by ice in June 1881, and most of its twenty-nine crew members perished before reaching Siberian settlements.

V

VACUUM TECHNOLOGY. Edison's development of an effective vacuum pump was key to the success of his **electric lamp**. He was not the first inventor to try to protect an incandescent conductor from the oxidizing (burning) effects of the atmosphere, but the mechanical air pumps of the time were inadequate. Edison looked for stronger pumps in early 1879 as he began investigating the effects of tiny gas bubbles on **platinum** wire at high temperatures. That spring, he designed a hybrid of two complementary types of pumps available in a few university laboratories. These instruments used mercury—a liquid at ordinary temperatures—to drive air from an enclosed space. Edison then contracted with **New York City** glassblowers to make the complex devices. After the design proved its merits (by several orders of magnitude over mechanical pumps), he hired **Ludwig Böhm** as his resident glassblower. Thus, he was ready that fall, when experiments showed the promise of thin strands of carbon, a material with the electrical properties he desired but one given to oxidizing in trace amounts of oxygen. **Francis Jehl** and **Francis Upton** refined the design, and by the summer of 1880, Edison was ready to make and install hundreds of pumps in his lamp factory. Evacuating lamps required repeatedly lifting mercury to recycle it through the pumps, a hard manual job in the laboratory made easy by electric motors in the factory. However, there was no easy way to prevent the neurotoxic effects of handling mercury. Although Edison discounted the hazards, there are strong suggestions (despite the lack of records from the **Edison Lamp Company**) of illness attributable to vacuum pumps at the factory.

In 1881, Edison's British agent **George Gouraud** encouraged him to try preserving perishable food in a vacuum to help meet Great Britain's growing import demands. Edison made several attempts, but when samples reached London, Gouraud found the fresh fruit decomposed and the meat so "noxious" that he fled the room (*TAEB* 6:94 n. 2). Edison had more success with vacuum processes for depositing an exceedingly thin layer of a vaporized metal such as gold or platinum. The idea was not entirely new when he tried to patent it in 1884, but he was able to claim originality in 1887 for using the process to duplicate wax **phonograph** recordings. Edison turned to this process in the early 1900s to prepare masters from which duplicates (Edison "gold molded" cylinders) were made on a large scale. Similar vacuum deposition processes later became important in unrelated industries, including semiconductors.

VAGABONDS. A group consisting of Edison, **John Burroughs**, **Harvey Firestone**, and **Henry Ford**, the self-styled Vagabonds took well-publicized summertime automotive camping trips. The idea originated in 1914, when Ford and Burroughs visited Edison in Florida and the trio drove through the Everglades. In 1915, Edison, Ford, and Firestone drove from Riverside to San Diego, California, following their attendance at the Panama-Pacific Exposition in San Francisco. The next year, Edison asked Ford, Burroughs, and Firestone to join him in the Adirondack and Green mountains (Ford did not go). The core group traveled together annually, often through portions of the Appalachian Mountain chain but also to Michigan on several occasions. After Burroughs died in 1921, the surviving troika carried on. The Vagabonds' trips were publicity bonanzas, to the delight of newspaper reporters and the advertising departments of the Ford Motor and Firestone Tire and Rubber companies. The 1919 entourage encompassed fifty vehicles, two of which Ford created for the occasion: a mobile kitchen with a gasoline stove and icebox and a heavy touring car with equipment including electric lights. In 1921, the group (sans Burroughs) briefly welcomed President Warren Harding in the Blue Ridge Mountains; three years later, they descended on President Calvin Coolidge at his Vermont summer home. Finally overwhelmed by the attention and publicity they created, the Vagabonds discontinued their travels after 1924.

VICTOR TALKING MACHINE COMPANY.

The firm was a dominant force in the recording industry and the chief competitor of Edison's **National Phonograph Company** in the early twentieth century. **Emile Berliner** and Eldridge Johnson, a machinist who had improved the fidelity of Berliner's **gramophone** recordings (in part by melting Edison cylinders to use the wax in his own masters), founded Victor in 1901. They combined their patents and manufacturing operations for records and instruments, with Johnson as president and Berliner a major stockholder. Victor built and successively enlarged a factory complex in Camden, New Jersey, to assemble gramophones and press discs from recordings made in nearby Philadelphia. It had emerged as a strong competitor to National Phonograph by 1907, when an economic depression hurt both companies' sales. National struggled afterward, but Victor sales rebounded strongly. Its recovery was due in part to the superior length (four minutes), cost, compactness, and durability of its discs against Edison cylinders, but there was also a sociological dimension. National

had focused on marketing popular artists that sold well in small towns and rural areas—regions particularly affected by the Great Depression. Victor's broader catalog included classical and operatic music that sold well in urban areas whose affluent residents had weathered the crisis better. Victor also offered a wider price range of machines and, as a result, found itself barely able to meet demand for high-end instruments. The company introduced the Victrola—a gramophone enclosed in a cabinet—in 1906, and its entire line of Victrolas proved very popular. To compete, Edison developed his own line of cabinet-enclosed phonographs known as Amberolas. The Radio Corporation of America (RCA) bought out Victor in 1929.

VILLAGE PLANT SYSTEM. Starting in 1882, Edison designed an alternative electrical distribution system for small cities and towns whose relatively low population density could not justify the costs of building and operating a system modeled on his **Pearl Street Station ("First District")** plant in **New York City**. A distinctive feature of the so-called village plant was the relatively high voltage of its generation and distribution (330 volts initially and 200 volts later, compared with 110 in New York), which allowed for thinner, less expensive wires and wasted less energy in transmission. The village plant model also marked a shift from big dynamos coupled directly to steam engines to smaller machines easier to design and build for the higher speeds enabled by belt drives. Edison's first village plant went into service in Roselle, New Jersey, in January 1883. One quirk of the design was that each branch circuit had several lamps that went on or off together. Dissatisfied with that feature, Edison created the **three-wire system**, in which each lamp could operate independently.

VILLARD, HENRY (1835–1900). A journalist, transportation entrepreneur, and financier, Villard played key roles in Edison's light and power enterprises and electric railway research. Born Heinrich Hilgard in Bavaria, he changed his name after immigrating to the

United States in 1853, in part so that his family could not locate him. Villard distinguished himself as a newspaper correspondent during the Civil War. After securing his initial fortune as an agent for German bondholders in American companies in the 1870s, he invested in American railroads, particularly in the Pacific Northwest. He returned to journalism in 1881 as publisher of the *New York Evening Post* and *The Nation*.

As the head of a rail and maritime empire in the Northwest, in 1880, Villard commissioned Edison to install electric lights on the steamship *Columbia*, the first ship so equipped. He served as a director of the **Edison Electric Light Company** and, in 1884, chaired its committee on manufacturing and reorganization. Bankruptcy compelled him to resign both posts. He moved to Berlin and helped align the interests of the French and German Edison companies to promote construction of central stations. Returning to **New York City** in 1886 as a representative of Deutsche Bank and as an adviser to German investors, he helped organize the Allgemeine Elektricität-Gesellschaft, the German Edison electric light and power company, in 1887. He brought in German investors to recapitalize and reorganize the Edison lighting interests as the **Edison General Electric Company (EGE)**, becoming EGE's president. Villard began to seek a merger with the **Thomson-Houston Electric Company**, but the firms' eventual consolidation (arranged by **Drexel, Morgan & Company**) into the **General Electric Company** in 1892 came over his objections, and he was forced out of the new company.

Villard controlled the Northern Pacific Railway for several years until his bankruptcy, and in 1881, he began encouraging and backing Edison's electric railway research. Becoming interested in urban streetcar lines in the late 1880s, he renewed his support of Edison's electrification efforts and funded a new test track at the **laboratory** in **West Orange, New Jersey**. He hoped to create a unified system for electric traction, light, and power in major cities, but his exclusion—and Edison's withdrawal—from General Electric ended that project.

VITASCOPE. Soon after Edison introduced the peephole **kinetoscope** for viewing **motion pictures**, inventors in Europe and the United States began working on projection systems to accommodate more than one viewer at a time. The **Lumière Brothers** introduced the first commercial projection system in Paris in 1895. That same year, **William K. L. Dickson** left Edison's employ to work on a projector with Woodville Latham and his sons, who exhibited what they called an eidoloscope. At the Cotton States Exposition in Atlanta, Thomas Armat and his partner C. Francis Jenkins exhibited their phantoscope. Frustrated by slowing kinetoscope sales and anxious to revitalize their enterprise, Kinetoscope Company agents Norman Raff and Frank Gammon asked Edison to develop a projection system. However, Edison's own interests were tied to kinetoscope sales, and he worked only half-heartedly on projection. Raff and Gammon then went around him and acquired rights to Armat's phantoscope for the Kinetoscope Company. Edison, recognizing competition from film companies using projection systems, agreed to have the **Edison Manufacturing Company** make phantoscope projectors and films for the Kinetoscope Company, which branded them as the Edison Vitascope. The Vitascope was introduced commercially in 1896 without mention of Armat's inventive role.

VOTE RECORDER. Edison's first patented invention was a vote recorder for use by legislative bodies such as Congress. He may have been spurred by reports in *The Telegrapher* that the Washington, D.C., city council planned to install an electric vote recorder and that the New York state legislature was considering one as well. Edison received U.S. Patent 90,646 for his device on 1 June 1869. The patent described a system in which each legislator moved a switch to either a yes or a no position to transmit a signal to a central recorder where legislators' names appeared in two columns of metal type headed "Yes" and "No." Dials on either side of the machine recorded the total number of yeas and nays, while a record of the individual votes was printed on paper by electrochemical decomposition in a manner similar to Edison's **automatic telegraph**. A fellow telegrapher named Dewitt Roberts bought an interest in the invention for $100. Edison later recalled that Roberts took it to Washington, D.C., to exhibit to a committee of Congress, whose chairman told him that "if there is any invention on earth that we don't want down here, that is it" (Dyer and Martin 1910, 103). Because legislators used the slow pace of roll call voting to filibuster or convince others to change their minds, Edison's instrument was never used.

W

WANGEMANN, ADELBERT THEODOR EDUARD (1855–1906). Possibly the world's first recording engineer, Wangemann was responsible for musical recordings at Edison's **laboratory** from 1888 until 1890. He also exhibited the **phonograph** to influential audiences in Europe. Wangemann was a talented pianist who acquired some education in engineering and music before emigrating from Berlin in 1879. He worked in the paper trade in **Boston**, moved to **New York City**, became a naturalized citizen, and started at Edison's laboratory in the spring of 1888. At first, Wangemann did experimental work on the phonograph, but by September, he was in charge of musical recordings. In early 1889, he selected artists and made musical and spoken-word recordings for demonstrating the new wax-cylinder phonograph. That spring, he recorded 880 musical cylinders, creating duplicates by recording one performance on multiple phonographs. In June, he took 654 cylinders to Paris for the **Paris Universal Exposition (1889)**. Wangemann accompanied Edison and his wife through the Exposition and recorded portions of a reception for them atop the Eiffel Tower. After making other recordings in Paris, he toured Europe with the phonograph and captured the voices of prominent figures such as Otto von Bismarck, Johannes Brahms, and William Gladstone.

Wangemann returned to the United States in early 1890 to find that Edison, still without a good way of duplicating cylinders, had given up making musical recordings at the laboratory. Wangemann supervised the phonograph exhibit at an industrial fair in Minneapolis in mid-1890 and then took a job with the New York Phonograph Company. He returned to Edison's laboratory in 1902 and again took up phonograph research. He also served on a special committee charged with reviewing recordings before their public release and as secretary of the Muckers of the Edison Laboratory, a fraternal organization of research assistants. Wangemann suffered fatal injuries while attempting to board a moving train.

WAR OF THE CURRENTS. Also called the Battle of the Currents or the War/Battle of the Systems, this series of events was a contest in the late 1880s and early 1890s between **direct current (DC)** and **alternating current (AC)** systems for generating and distributing electricity. Partly a war of words, it was often oversimplified as a dispute between Edison and **George Westinghouse**. The personal animus was real (and made good newspaper copy), and partisans such as **Harold Brown** turned genuine physiological and moral questions about **electrocution** for animal euthanasia and capital punishment into polemics. Yet in essence, the "war" was a matter of technology and economics in what author Jill Jonnes called a "race to electrify the world," starting with the United States (Jonnes 2003). Edison had a head start with DC, and his early success created a huge demand for generation and distribution capacity. Ironically, the large systems needed to meet those demands were cheaper to build and operate with AC than with DC. Edison tried to invent his way to a more economical

DC system, mainly by driving up the efficiency of **electric lamps** and transmitting current at higher voltages. He also made some belated attempts to develop AC generators and transformers. (The **Edison Electric Light Company** bought U.S. patent rights to a promising AC system, apparently to bury rather than develop it.) Concerned about safety, Edison opposed the high pressures (above 1,000 volts) used in AC systems—as did politicians in a growing number of cities after gruesome accidental deaths involving AC. The 1892 merger of the **Edison General Electric Company** with rival **Thomson-Houston Electric Company** (which had patents for both AC and DC systems) symbolized the defeat of DC and the end of the "war." However, DC systems remained common in the United States for decades, integrated into the growing AC transmission grid by converter "substations." Local utilities began a general changeover from DC to AC in the 1920s, though the last DC station in **New York City** was decommissioned in 2007. Recent years have seen the construction of high-voltage DC systems in place of AC for long-distance transmission.

WEEKLY HERALD. While working as a newsboy for the Grand Trunk Railway in early 1862, fourteen-year-old Edison bought 300 pounds of old type slugs and a small second-hand press. After installing the equipment (and a chemistry **laboratory**) in an unused cubby in the baggage car of his regular train, he debuted the one-page *Weekly Herald* on 3 February 1862. This first issue included news from towns along the line, moral instruction à la Benjamin Franklin, humor, and coach schedules from Grand Trunk stations. The publisher ("A. Edison") promised larger papers in the future with the names of all subscribers to be printed (*TAEB* 1:25–26). His experience selling established newspapers along the line surely raised the young Edison's commercial hopes. His enterprise enlisted as many as 500 subscribers at eight cents per month, but a fire in the rolling chemistry lab terminated Edison's onboard publishing privileges after about six months. There is some evidence that he continued to publish the *Weekly Herald* from

his home in **Port Huron, Michigan**, where he and an associate later put out a disreputable and short-lived gossip sheet, *Paul Pry. See also* CIVIL WAR TELEGRAPHY.

WEST ORANGE, NEW JERSEY. Edison lived more than half his life in West Orange, where he had his second family and built his largest **laboratory**. The city was part of Newark township (which encompassed all the present-day Oranges) until November 1806. Orange was incorporated as a town in 1860 and reincorporated as a city in 1872. In that interval, it began fragmenting into smaller communities, primarily because of local disputes about the cost of establishing police, fire, and municipal departments. West Orange was incorporated as a separate township in April 1863 and reorganized as a town in 1900. It included the private community of **Llewellyn Park**, where, in 1886, Edison bought **Glenmont**, a three-story mansion on thirteen and a half acres. He built a new laboratory the next year—and eventually an entire factory complex—on Main Street, within walking distance of the home.

WESTERN ELECTRIC MANUFACTURING COMPANY. This Chicago-based firm grew from Gray & Barton, a Cleveland manufacturing partnership formed in 1869 by inventor Elisha Gray (of later telephone fame), Enos Barton (a former **Western Union Telegraph Company** manager), and Anson Stager (a wealthy Western Union divisional superintendent). At the urging of Western Union, which needed manufacturing capacity in the growing western states, Gray & Barton relocated to Chicago in 1870. Two years later, it sold a one-third ownership interest to Western Union and changed its name to Western Electric. Although closely aligned with Western Union, Western Electric was free to take on other work, and in 1876, it contracted with Edison to manufacture and sell the **electric pen** copying system. That arrangement allowed Edison to reach a much larger market and provided him with significant income at a crucial moment in his young career. It also brought him into contact with **George Bliss**, who became a useful promoter of Edison inventions. A few years later, Western Electric began making telephones for Western Union subsidiary **Gold and Stock Telegraph Company** under the patents of Edison and Gray. (Edison supplied the carbon buttons for the transmitter from **Menlo Park, New Jersey.**) Edison also made Western Electric his agent for the telephone in Australia. Western Union's withdrawal from the telephone business in 1879 deprived Western Electric of a large portion of its business. In response, Western Electric acquired the Indianapolis factory of **Ezra Gilliland**—a major contractor to the **American Bell Telephone Company**—and leveraged its position into becoming sole supplier to American Bell and subsequently the vast Bell system.

WESTERN UNION TELEGRAPH COMPANY.

Founded in 1851 as the New York and Mississippi Valley Printing Telegraph Company, Western Union was renamed five years later after a merger. It acquired its major rivals after the Civil War (during which Edison worked as a Western Union operator) and achieved virtual monopoly control of the American telegraph industry by the early 1880s. In the 1870s, Edison became a contract inventor for the company as well as for the competing **Automatic Telegraph Company** And the **Atlantic and Pacific Telegraph Company**. In 1871, Western Union acquired the **Gold and Stock Telegraph Company** (which had just retained Edison as its consulting electrician), getting his key printing telegraph patents in the bargain. Western Union also funded Edison's 1873–1874 experiments on the **duplex telegraph** and the **quadruplex telegraph**. Yet its failure to ensure control of the quadruplex patents allowed Edison to sell them to **Jay Gould**'s rival Atlantic and Pacific Telegraph Company, leading to a long legal contest. Although perturbed by Edison's sale of quadruplex rights, Western Union president **William Orton** agreed to support his experiments on **acoustic telegraphs** and telephone technology. Norvin Green became president after Orton's death in 1878, and the next year, he sold Western Union's telephone patents to the **American Bell Telephone Company**. By that time, Edison had turned to electric lighting,

with investors connected to Western Union providing most of the initial funds. Edison did his subsequent work on the **railway ("grasshopper") telegraph** independently of Western Union, and American Bell prompted his later telephone improvements.

WESTINGHOUSE ELECTRIC COMPANY.

This Pittsburgh, Pennsylvania, company was created in January 1886 by **George Westinghouse** to develop, manufacture, and sell an **alternating current** (AC) electric light and power system. Talented engineers such as **Nikola Tesla**, William Stanley, and Oliver Shallenberger developed **electric motors**, transformers, and meters. However, the company did not hesitate to acquire useful inventions from the outside, and its purchase of patent rights to an early transformer gave it control over the fundamental principles of an AC distribution system. Westinghouse Electric focused on incandescent lighting and competed directly with the **Edison Electric Light Company**, exploiting AC's advantages in relatively low-density central station districts where the cost of copper conductors for an Edison-style **direct current** system was prohibitive. Its cash-only plan for building central stations (the Edison model relied on stocks and bonds) helped it meet the capital needs of a rapidly expanding business. By 1891, as Westinghouse neared victory in the **War of the Currents**, the number of its central stations and their generating capacities approached those of the **Edison General Electric Company** (EGE) and exceeded those of the **Thomson-Houston Electric Company**.

For a time in the late 1880s, Westinghouse Electric made its own lamps under the patents of **William Sawyer**. When federal courts voided Sawyer's key patent in 1892 in favor of Edison's sealed high-vacuum bulb, Westinghouse Electric was ready with a fifty-volt "stopper" lamp closed off by cementing a glass stopper into the bulb, like a cork in a bottle. Although not as durable as the Edison lamp, the stopper lamp evaded the basic Edison patent, sufficient for the company's immediate purposes. Westinghouse Electric used it to light the **World's Columbian Exposition**

(1893), a brilliant success that led to a contract to supply the equipment for the first large-scale hydroelectric generating plant at Niagara Falls (1895). Rumors of a merger between Westinghouse and EGE (and its successor, **General Electric Company**) circulated off and on for years. Instead, with hundreds of lawsuits pending between them, Westinghouse Electric and General Electric agreed in 1896 to cross-license patents to each other, allowing each side to make and sell a full range of equipment and to compete mainly on price. The company entered receivership after the Panic of 1907 but recovered and grew into a manufacturing colossus. *See also* ELECTROCUTION (ANIMAL EXPERIMENTS); ELECTROCUTION (CAPITAL PUNISHMENT).

WESTINGHOUSE, GEORGE (1846–1914). American inventor and manufacturer based in Pittsburgh, Pennsylvania, Westinghouse developed and promoted an **alternating current** (AC) electric system to rival Edison's **direct current** (DC) and served as the public face of AC in the **War of the Currents**. He entered the electric lighting field relatively late, in 1884, after making his reputation and fortune in railroad airbrake and signal systems. He had also designed a delivery network for natural gas that reduced the high pressure at the wellhead to a safe level for homes and factories—a scheme that became a template for a high- and low-pressure network for electrical energy. In addition to his own inventive abilities, Westinghouse had talent for improving the inventions of others and for hiring well. Starting with an early transformer design and the skill of electrical engineer William Stanley, he created the basics of a complete AC distribution system by late 1885. He founded the **Westinghouse Electric Company** to bring it to market.

Westinghouse's relative lack of electrical experience drew Edison's scorn ("the man has gone crazy . . . and is flying a kite that will land him in the mud sooner or later" [*TAEB* 9:522]), but his rapid success commanded attention ("He is ubiquitous and will form innumerable companies before we know anything about it" [*TAEB* 8:640]). Westinghouse became the

target of harsh attacks by Edison companies and their surrogates. Before the adoption of the word "electrocute," "westinghouse" was suggested as a suitable verb, and Edison was quoted in a newspaper calling his rival a "shyster" (*TAEB* 9:900). Edison also refused an invitation to visit Pittsburgh and attempted to undermine Westinghouse's signature industrial enterprise by developing an electric railroad brake. (The two men did appear in the same Pittsburgh courtroom in May 1889 for the start of a high-stakes patent lawsuit.) For his part, Westinghouse demurred when **Henry Villard** floated the idea of combining the Westinghouse and Edison electric companies. He gave up full control of Westinghouse Electric when the Panic of 1907 forced it into receivership, but he remained involved with the company until he retired in 1911.

WESTON, EDWARD (1850–1936). Englishman Edward Weston was a successful electroplater, electrical inventor, and manufacturer in the United States whose foray into electric lighting ignited a sharp rivalry with Edison. After launching his electroplating company in **New York City**, Weston moved to **Newark, New Jersey**, where he became one of the nation's leading dynamo builders. He entered the arc lighting business in 1879 on the strength of his patented lamps. When *Scientific American* featured Edison's new bipolar **electric generator** in October of that year, Weston responded that Edison's claims for the machine were "mathematically absurd" and proof of Edison's ignorance (*TAEB* 5:451 n. 3). Edison replied in kind, asserting that Weston's "statements are without sense or science, and plainly originate from one who does not understand the laws which he pretends to set forth" (*TAEB* 5:449). In fact, Weston was expressing a widespread misunderstanding about the relationship between a dynamo's internal resistance and that of the outside circuit, presuming that it should be designed for maximum output (for electroplating), whereas Edison aimed to optimize efficiency (for lighting). Weston branched into incandescent lighting, and his celluloid filament was a major advancement, though Edison always saw him as an interloper and among the "pirates" who were "constantly copying what I do" (*TAEB* 7:156). In the early 1880s, Weston joined the United States Electric Light Company, which became a principal competitor to the Edison interests and won the contract to illuminate the new Brooklyn Bridge. He built a personal laboratory in Newark, described in 1887 as "probably the most complete in the world." That attainment surely galled Edison and perhaps goaded him to plan what became his new **laboratory** in **West Orange, New Jersey**. Weston left the electric lighting business and produced instruments for making electrical measurements.

WILBER, ZENAS FISK (1839–1889). Wilber was a Patent Office examiner who became Edison's principal patent attorney in 1880 (partly through the intercession of **Josiah Reiff**). He formed a partnership in Washington, D.C., with George Dyer; personally handled Edison's new applications; and was on retainer for the Edison Electric Light Company. When Dyer & Wilber dissolved in 1882, Edison named **Richard Dyer** (George's son and a **New York City** resident) as his principal attorney. Wilber filed applications prepared by the younger Dyer and continued on retainer for the company. Those arrangements lasted about half a year until it became clear that Wilber had failed to file scores of applications while pocketing the fees. Edison later claimed to have lost seventy-eight patents to Wilber's malfeasance. In 1886, Wilber admitted to acting corruptly at the Patent Office in an even more consequential matter ten years earlier: awarding priority of invention for the telephone to **Alexander Graham Bell** over the rival claims of **Elisha Gray**. (He was heavily in debt to one of Bell's attorneys.) His reputation ruined, Wilber succumbed to acute alcoholism ahead of giving much-anticipated testimony in a federal lawsuit about the telephone patents.

WIZARD OF MENLO PARK. This sobriquet was Edison's most enduring nickname, bestowed in April 1878 by the *New York Daily Graphic*. The phrase headlined an article by William Croffut, one of many reporters from **New York City** and elsewhere who descended

The *New York Daily Graphic*, which bestowed the nickname on Edison in 1878, depicted the "Wizard's Search" for platinum the next year. *Courtesy of Thomas A. Edison Papers.*

on the **laboratory** in **Menlo Park, New Jersey**, in March and April to see the **phonograph** and describe its inventor for their readers. The phonograph so astounded both the scientific community and the public that Edison seemed capable of inventing anything. In fact, Croffut had published not two weeks earlier an April Fool's story claiming that Edison had created a machine to manufacture food out of air, water, and earth. Other reporters described Edison as the "Napoleon of Invention" and the "Inventor of the Age," but it was the "Wizard of Menlo Park" moniker that stuck for the rest of his life. *See also* PLATINUM; PRESS RELATIONS.

WORLD WAR I RESEARCH. In mid-1915, following the attack on the steamship *Lusitania* but with hopes still alive that the United States could avoid the Great War in Europe, Edison put forward a plan of national preparedness in case it could not. His proposal was based on raising a corps of engineers in civilian industries and on inventing and creating a vast stockpile of advanced military equipment, such as airplanes and submarines. The plan ultimately contributed to the formation of the **Naval Consulting Board**. By the time the United States declared war in April 1917, Edison had already turned his research and development efforts to experiments for the U.S. Navy.

Working at his **laboratory** in **West Orange, New Jersey**; in Key West, Florida; and in the waters off Sandy Hook, New Jersey, and Long Island, he designed means of locating enemy submarines, airplanes, and torpedoes; a submarine searchlight; a wireless telegraph scrambler; cannon-fired steel mesh drapes to slow incoming torpedoes; and a water brake to quickly turn ships. Edison financed some of the projects himself, his only perk being the USS *Sachem*, a 210-foot yacht leased from the navy. According to later tallies, he made about fifty-seven inventions, plans, and designs for the navy, and Edison complained that "they pigeonholed every one of them" (Morris 2019, 210). Among them was a plan to conduct merchant shipping by night in designated lanes to avoid German U-boat attacks. The idea cost several months of research in cartography and navigation at Washington, D.C., and the navy's dismissal of it (by "mentally inbred" officers) dumbfounded Edison (Morris 2019, 21). By the time he shifted his research to the garage at **Glenmont** in the spring of 1918, his relationship with the naval brass was crumbling. Yet he continued his work even as the war began to close, putting in for a ship larger than the *Sachem*. Secretary of the Navy **Josephus Daniels** denied it, and the navy soon withdrew the *Sachem* itself from service. Edison carried on until September 1919, ten months after the Armistice, when Assistant Secretary of the Navy Franklin Roosevelt told him that his efforts were no longer needed.

WORLD'S COLUMBIAN EXPOSITION (1893).

The Exposition was a world's fair held in Chicago to celebrate the 400th anniversary of Christopher Columbus's arrival in the New World. Open from 1 May through 31 October, the affair garnered 21.5 million paid admissions to popular attractions such as the original Ferris wheel, German beer halls, Egyptian jugglers, a 1,500-pound chocolate Venus de Milo, and the World's Congress of Beauty. As early as the spring of 1891, Edison declared his intent to display his **kinetoscope** at the Exposition, yet in November 1892, **Alfred Tate** was still urging him "to take this [kinetoscope] matter in hand at once" (Spehr 2008, 273). By that time, Edison was focusing on the Ogden ore milling plant, and he formally surrendered the kinetograph concession less than a month before the fair opened. Another opportunity was lost when the **Westinghouse Electric Company** won the contract to light the Exposition's buildings and grounds by underbidding the **General Electric Company** (successor to the **Edison General Electric Company**) by more than $600,000. The Westinghouse forces also mounted an exhibit showcasing their **alternating current** distribution system. Edison, remembering his losses incurred at the **Paris Universal Exposition (1889)**, chose not to have an electrical display at Chicago, although he did stage a historical exhibit (including some **phonographs** and an **electric generator** from the **Pearl Street Station** ["First District"]) via the **North American Phonograph Company**. Edison attended the Exposition, noting that "it is all marvelous," but the event was never a priority for him. *See also* WAR OF THE CURRENTS.

X

X-RAYS. Edison was one of the many electrical inventors, engineers, and scientists who investigated X-rays after Wilhelm Röntgen discovered them in December 1895. Edison's primary aim was to develop practical apparatus for the study and use of this new phenomenon. He and his staff experimented with dozens of different materials and shapes for the vacuum tubes and electrodes used to generate X-rays, tested different power sources and spark generators, and explored how readily the rays penetrated different materials. They also investigated nearly 1,300 compounds to find the best fluorescing salt for use in X-ray photographic plates and screens for medical uses.

This research led to the development of the calcium-tungstate screen fluoroscope, which Edison refused to patent and arranged to manufacture and sell at a modest price. Although the fluoroscope allowed doctors to view an X-ray image immediately, they preferred the permanent record of the X-ray photograph. Edison's primary assistant in his X-ray work was glassblower Clarence Dally, who died in 1904 from the effects of radiation damage. In a futile effort to save his life from cancer, Dally had his left hand, four fingers on his right hand, and finally both arms amputated. Dally's illness and death led Edison to give up X-ray research.

Bibliography

CONTENTS

INTRODUCTION

Compiling an Edison bibliography is a challenging task. Lists of contemporaneous and modern biographies (including significant juvenile works) are relatively easy to assemble, but Edison's own publications and the massive secondary literature about him require more diligence and discretion. As famous as Edison was in his lifetime, his words (or those ascribed to him) were endlessly reprinted by newspapers and magazines. And with his work having influenced so many technologies and businesses, the historical literature around his life and career is wide and deep.

This bibliography has four sections. We begin with a short list of major archival collections, although Edison materials can be found in hundreds of repositories around the globe. Then we provide a bibliography of Edison's own publications. The first part of this section encompasses Posthumous Collections, including those produced by the Thomas A. Edison Papers at Rutgers University. The longer second subsection comprises Edison's own Published Articles and Select Interviews (listed by year) in American magazines and journals. Early in his career, Edison published extensively in telegraph industry journals *The Telegrapher*, the *Journal of the Telegraph*, and the *Operator*, and for a time, he was science editor of the latter. He also contributed articles (some as letters to the editor) in *Scientific American* and reported his chemical research in the *American Chemist* and *Chemical News*. Between 1878 and 1888, he prepared papers for annual meetings of the American Association for the Advancement of Science, which published them in its *Proceedings*. After the phonograph made Edison famous in 1878, the *North American Review* and other American periodicals occasionally solicited contributions from him. During the 1890s, he published in *Electrical Engineer*, particularly about his X-ray research. Beginning in the 1890s and especially in the 1910s and 1920s, several periodicals ran extensive interviews with him (some even under his byline). Because these interviews provide depth to Edison's views on a range of subjects, including broader social issues, we have chosen to include them with his publications. Newspaper interviews are generally excluded because of the special problems they present. American papers printed innumerable interviews and articles based on conversations with him. Other papers often picked up these stories, reprinting them in whole or in part, sometimes months later, making the modern bibliographer's job that much harder without a commensurate payoff.

The bibliography's third section covers biographical works and is divided into three subsections: (1) Modern Biographies, (2) Contemporaneous Biographies and Sketches, and (3) Select Juvenile Works.

The final—and largest—section contains thirteen subject areas related to Edison's work and influence. These categories are (1) Electric Light and Power; (2) Phonograph and Sound Recording; (3) Motion Pictures; (4) Telecommunications; (5) Batteries and Electric Automobiles; (6) Materials Processing (Ore Milling, Cement, Chemicals, and Rubber); (7) World War I; (8) Miscellaneous Topics; (9) Laboratories, Patents, and Research Methods; (10) Entrepreneurship and Economic Ideas; (11) Edison Myth and Reputation; (12) Edison Family and Fort Myers; and (13) Edison Associates and Rivals.

Archives and Papers Collections

Charles Edison Fund. http://www.charles edisonfund.org

Edison Collection. Benson Ford Research Center. The Henry Ford. https://findingaids.the henryford.org/xtf/data/pdf/ThomasA EdisonCollection_Accession1630.pdf

Thomas A. Edison Papers Digital Edition (*TAED*). https://edisondigital.rutgers.edu

Thomas Edison National Historical Park archives. https://www.nps.gov/edis/learn /historyculture/research.htm

Publications by Thomas A. Edison

1. Posthumously Published Collections

Jeffrey, Thomas E., et al. *Thomas A. Edison Papers: A Selective Microfilm Edition.* Frederick, MD: University Publications of America, 1985–1999; LexisNexus, 2008.

Jenkins, Reese V., Robert Rosenberg, Paul Israel, et al. *The Papers of Thomas A. Edison.* Vols. 1–10 (*TAEB*). Baltimore: Johns Hopkins University Press, 1989–. [Includes autobiographical reminiscences.] https://edison.rutgers.edu/research/book -edition

McGuirk, Kathleen L., ed. *The Diary of Thomas A. Edison.* New York: Chatham Press, 1971.

Rosenberg, Robert, Paul Israel, Thomas E. Jeffrey, et al. *Thomas A. Edison Papers Digital Edition.* 2000–. https://edisondigital .rutgers.edu

Runes, Dagobert, ed. *The Diary and Sundry Reminiscences of Thomas Alva Edison.* New York: Philosophical Library, 1948.

2. Authored Articles and Select Interviews

[1868]

Editorial Notice. *Telegrapher* 4 (11 January 1868): 163.

Editorial Notice. *Telegrapher* 4 (4 April 1868): 258.

"Edison's Double Transmitter." *Telegrapher* 4 (11 April 1868): 265.

"The Induction Relay: To the Editor." *Telegrapher* 4 (25 April 1868): 282.

"Edison's Combination Repeater." *Telegrapher* 4 (9 May 1868): 298.

"Automatic Telegraphing" [by M. F. Adams]. *Journal of the Telegraph* 1 (1 June 1868): 3.

"To the Editor." *Telegrapher* 4 (2 June 1868): 334.

"Self-Adjusting Relays." *Telegrapher* 4 (8 August 1868): 405.

"The Manufacture of Electrical Apparatus in Boston." *Telegrapher* 4 (15 August 1868): 413–14.

"American Compound Telegraph Wire." *Telegrapher* 5 (17 October 1868): 61.

[1869]

Editorial Notice. *Telegrapher* 5 (30 January 1869): 183.

"A New Double Transmitter" [editorial notice]. *Telegrapher* 5 (17 April 1869): 272.

"Pope, Edison & Co. Advertisement." *Telegrapher* 6 (2 October 1869): 45.

"Queries: To the Editor." *Telegrapher* 6 (16 October 1869): 58.

"Edison's Button Repeater." In *Modern Practice of the Electric Telegraph*, by Frank L. Pope, 107–8. New York: D. Van Nostrand, 1869.

[1874]

[Edison's quadruplex telegraph.] *Operator* (15 July 1874): 5.

"Duplex Telegraphy." Part 1 of 3. *Operator* (1 September 1874): 1.

"To the Editor." In "The Electromotograph—A New Discovery in Telegraphy." *Scientific American* 31 (5 September 1874): 145.

"Platina Points: To the Editor." *Operator, Supplement* (15 September 1874): 2.

"Duplex—No. II." Part 2 of 3. *Operator* (1 October 1874): 1.

"On a New Form of Relay." *Telegraphic Journal* 2 (1 October 1874): 319–20.

"Cable Telegraphy: To the Editor." *Scientific American* 31 (7 November 1874): 292.

"Duplex—No. III." Part 3 of 3. *Operator* (15 November 1874): 1.

"On a New Method of Working Polarised Relays." *Telegraphic Journal* 2 (15 November 1874): 361.

"To the Editor." *Operator* (1 December 1874): 3.

"Cable Telegraphy: To the Editor." *Scientific American* 31 (12 December 1874): 372.

[1875]

"On the Imperfect Contacts Which Occur in Signalling with Rigid Contact-Points." *Journal of the Society of Telegraph Engineers* 4 (1875): 117–19.

[1876]

"To the Editor." *Scientific American* 34 (1 January 1876): 2.

"Mr. Edison's New Force: To the Editor." *Scientific American* 34 (5 February 1876): 81.

"Mr. Edison's New Force: To the Editor." *Scientific American* 34 (12 February 1876): 101.

"Laboratory Notes," nos. 1–7. *American Chemist* 7 (October 1876): 127.

[1877]

"Laboratory Notes," nos. 8–11. *American Chemist* 7 (March 1877): 356.

"Edison's Pressure Relay." *Journal of the Telegraph* 10 (1 June 1877): 163.

"To the Editor" [from Edward H. Johnson]. *Scientific American* 37 (17 November 1877): 304.

[1878]

With Edward H. Johnson. "The Phonograph and Its Future." *North American Review* 126 (May–June 1878): 527–36.

"To the Editor." *New York Tribune* (8 June 1878): 5.

"Mr. Edison on the Microphone: To the Editor." *Scientific American* 39 (13 July 1878): 20.

"To the Editor." *New York Tribune* (15 July 1878): 5.

"Professor Edison's New Carbon Rheostat." *Scientific American* 39 (20 July 1878): 35.

"Telephonic Repeater: To the Editor." *Chemical News* 38 (26 July 1878): 45.

"On the Use of the Tasimeter for Measuring the Heat of the Stars and of the Sun's Corona." *Proceedings of the American Association for the Advancement of Science* 27 (August 1878; pub. 1879): 109–12.

"The Sonorous Voltameter." *Proceedings of the American Association for the Advancement of Science* 27 (August 1878; pub. 1879): 112.

"To the Editor." *Scientific American* 39 (28 September 1878): 196.

"Telephone Relay: To the Editor." *Chemical News* 38 (18 October 1878): 198.

[1879]

"Clerac's Tube: To the Editor." *Telegraphic Journal* 7 (15 April 1879): 131.

"On the Phenomena of Heating Metals in Vacuo by Means of an Electric Current." *Proceedings of the American Association for the Advancement of Science* 28 (August 1879; pub. 1880): 173–78.

"On a Resonant Tuning Fork." *Proceedings of the American Association for the Advancement of Science* 28 (August 1879; pub. 1880): 178.

"A Note from Mr. Edison. The Hughes Microphone and the Blake Transmitter." *Scientific American* 41 (6 December 1879): 360.

"Edison's Telephonic Researches." In *Speaking Telephone, Electric Light, and Other Recent Electrical Inventions*, by George B. Prescott, 218–34. New York: D. Appleton & Co., 1879.

[1880]

"The Success of the Electric Light." *North American Review* 131 (October 1880): 295–300.

"Telegraph." By Edison and others. *Appleton's Cyclopedia* 2 (1880): 849–59.

[1882]

"Description of the Edison Steam Dynamo." Coauthored by Charles T. Porter. *Journal of the Franklin Institute* 114 (July 1882): 1–12.

"How to Succeed as an Inventor." In *How to Succeed in Public Life . . . A Series of Essays*, ed. Lyman Abbott, 95–104. New York: G. P. Putnam's Sons, 1882.

[1885]

"Electricity: Man's Slave." *New York Tribune* (18 January 1885): 10.

[1886]

"The Air-Telegraph: System of Telegraphing to Trains and Ships." *North American Review* 142 (March 1886): 285–91.

[1887]

"On a Magnetic Bridge or Balance for Measuring Magnetic Conductivity." *Proceedings of the American Association for the Advancement of Science* 36 (August 1887; pub. 1888): 92–94.

"On a Pyromagnetic Dynamo: A Machine for Producing Electricity Directly from Fuel." *Proceedings of the American Association for the Advancement of Science* 36 (August 1887; pub. 1888): 94–98.

[1888]

"The Perfected Phonograph." *North American Review* 146 (June 1888): 641–50.

[1889]

"Mr. Edison and His Phonograph: To the Editor." *New York Tribune* (23 January 1889): 7.

"The Dangers of Electric Lighting." *North American Review* 149 (November 1889): 625–34.

"The Concentration of Iron-Ore." Coauthored by John Birkinbine. *Transactions of the American Institute of Mining Engineers* 17 (February 1889): 728–44. Paper presented to the American Institute of Mining Engineers in New York.

[1890]

"An Account of Some Experiments upon the Application of Electrical Endosmose to the Treatment of Gouty Concretion." *Telegraphic Journal and Electrical Review* 27 (22 August 1890): 213. Paper presented to the International Medical Congress in Berlin, Germany.

[1892]

"Insulation." *Electrical Engineer* 14 (13 July 1892): 34–35.

[1893]

"Intelligent Atoms." In "Panpsychism and Panbiotism." *The Monist* 3 (1893): 242–45.

[1895]

"Edison on Inventions. A Remarkable Interview with the Great Inventor." Rufus R. Wilson interview. *Monthly Illustrator and Home and Country* 11 (1895): 340–44.

"Introduction." In William K. L. Dickson and Antonia Dickson, *History of the Kinetograph, Kinetoscope, & Kinetophonograph*. N.p.: W. K. L. Dickson, 1895.

[1896]

"Experiments with Roentgen Rays." *Electrical Engineer* 21 (25 March 1896): 305.

"Further Experiments in Fluorescence under the Cathode Ray." *Electrical Engineer* 21 (1 April 1896): 340.

"Are Roentgen Ray Phenomena Due to Sound Waves?" *Electrical Engineer* 21 (8 April 1896): 353–54.

"Roentgen Ray Lamps and Other Experiments." *Electrical Engineer* 21 (15 April 1896): 378.

"A Card from Mr. Edison: To the Editor." *New York Journal* (18 April 1896).

"Influence of Temperature on X-Ray Effects." *Electrical Engineer* 21 (22 April 1896): 409–10.

"Photographing the Unseen: A Symposium on the Roentgen Rays." *Century Magazine* 52 (May 1896): 120–31. [Edison's contribution appears on p. 131.]

"Recent Roentgen Ray Observations." *Electrical Engineer* 22 (18 November 1896): 520.

"In the Deep of Time." Coauthored by George Parsons Lathrop. Serialized in American newspapers by the literary syndicate of Irving Bachellor. See, for example, *Milwaukee Sentinel* (12 December 1896): 1, 20; (20 December 1896): 21; (27 December 1896): 19; (3 January 1897): 18.

[1897]

"Fluorescing Salts." *Electrical Engineer* 23 (6 January 1897): 17.

"Introduction." In *A Complete Manual of the Edison Phonograph*, ed. George E. Tewksbury, 10–12. Newark, NJ: United States Phonograph Co., 1897.

[1898]

"Edison on the Incandescent Lamp: To the Editor." *Electrical Review* 32 (5 January 1898): 7.

"Mr. Edison Protests against Yellow Journalism: To the Editor." *New York Sun* (12 January 1898): 6.

"Do Lightning Rods Protect?" *Popular Science News* 32 (May 1898): 116.

[1901]

"The New Edison Storage Battery." Coauthored by Arthur E. Kennelly. In *Transactions of the American Institute of Electrical Engineers* 28 (1901): 219–48.

[1902]

"The Storage Battery and the Motor Car." *North American Review* 175 (July 1902): 1–4.

[1904]

"Beginnings of the Incandescent Lamp." *Scientific American Supplement* 57 (14 May 1904): 23711–12.

[1910]

"Inventions of the Future." *The Independent* 68 (6 January 1910): 15–18.

[1911]

"Thomas A. Edison on Immortality." Edward Marshall interview. *Columbian Magazine* 3 (January 1911): 603–12.

"The Wonderful New World Ahead of Us." Allan L. Benson interview. *Cosmopolitan* 50 (February 1911): 294–306.

"Impressions of European Industries." *Scientific American* 105 (18 November 1911): 445–45.

"Edison on Co-operation vs. Competition." *Manufacturer's Record* 60 (14 December 1911): 49–50.

"Edison on Invention and Inventors." Waldon P. Warren interview. *Century Magazine* 82 (1911) 415–19.

"Mr. Edison Says: Electricity and Machinery Can Make Household Drudgery a Thing of the Past." Allan L. Benson interview. *The Delineator* 77 (1911): 7, 67.

[1912]

"The Woman of the Future." Edward Marshall interview. *Good Housekeeping* 55 (October 1912): 436–44.

"Edison Says Germany Excels Us." Allan L. Benson interview. *The World To-Day* 21 (November 1911): 1356–60.

[1913]

"Give the Inventor a Fair Chance. *Leslie's Illustrated Weekly* (2 January 1913): [4?].

[1914]

"Today and Tomorrow." John R. McMahon interview. *Independent* 77 (5 January 1914): 24–27.

[1915]

"Edison's Plan for Preparedness." Edward Marshall interview. *New York Times* (30 May 1915): SM:6–7.

[1922]

A Proposed Amendment to the Federal Reserve Banking System: Plan and Notes. West Orange, NJ: Thomas A. Edison, 1922.

[1923]

"How I Would Double the Volume of a Business." Samuel Crowther interview. *System* 44 (September 1923): 265–68, 330–32.

[1925]

"Introduction." In *The Life and Works of Thomas Paine*, vol. 1, by William M. van der Weyde, vii–ix. New Rochelle, NY: Thomas Paine National Historical Association, 1925.

[1926]

"Machine-Made Freedom." Edward Marshall interview. *Forum* 76 (October 1926): 492–49.

"Has Man an Immortal Soul?" Edward Marshall interview. *Forum* 76 (November 1926): 641–50.

"Scientific City of the Future." Edward Marshall interview. *Forum* 76 (December 1926): 823–82.

[1927]

"Youth of To-day and To-morrow." Edward Marshall interview. *Forum* 77 (January 1927): 41–53.

Biographies

1. Modern Biographies

Baldwin, Neil. *Edison: Inventing the Century.* New York: Hyperion, 1995.

Clark, Ronald. *Edison: The Man Who Made the Future.* New York: Putnam, 1977.

Conot, Robert. *A Streak of Luck.* New York: Seaview Press, 1979.

DeGraaf, Leonard. *Edison and the Rise of Innovation.* New York: Sterling Signature, 2013.

Gitelman, Lisa, and Theresa M. Collins. *Thomas Edison and Modern America: A Brief History with Documents.* New York: Bedford Books/St. Martin's/Palgrave, 2002.

Israel, Paul. *Edison: A Life of Invention.* New York: John Wiley & Sons, 1998.

Josephson, Matthew. *Edison: A Biography.* New York: McGraw-Hill, 1959.

Melosi, Martin V. *Thomas A. Edison and the Modernization of America.* Glenview, IL: Scott, Foresman/Little, Brown Higher Education, 1990.

Morris, Edmund. *Edison.* New York: Random House, 2019.

Stross, Randall E. *The Wizard of Menlo Park: The Life and Times of Thomas Alva Edison.* New York: Crown Publishers, 2007.

Vanderbilt, Byron. *Thomas Edison, Chemist.* Washington, DC: American Chemical Society, 1971.

2. Contemporaneous Biographies and Sketches

Ballentine, Caroline Farrand. "The True Story of Edison's Childhood and Boyhood." *Michigan Magazine of History* 4 (1920): 168–92.

Bryan, George S. *Edison, The Man and His Works.* New York: A. A. Knopf, 1926.

Cooper, Frederic Taber. *Thomas A. Edison.* New York: Frederick A. Stokes Co., 1914.

Dickson, William Kennedy Laurie, and Antonia Dickson. *The Life and Inventions of Thomas Alva Edison.* New York: Thomas Y. Crowell & Co., 1894.

Dyer, Frank L., and Thomas C. Martin, with William H. Meadowcroft. *Edison: His Life and Inventions.* 2 vols. New York: Harper & Bros., 1910 [rev. ed. 1929].

Ford, Henry, and Samuel Crowther. *Edison as I Know Him.* New York: Cosmopolitan Book Corporation, 1930.

Jones, Francis Arthur. *Thomas Alva Edison: Sixty Years of an Inventor's Life.* New York: Thomas Y. Crowell & Co., 1907, 1908.

McClure, James Baird. *Edison and His Inventions.* Chicago: Rhodes & McClure, 1879, 1889, 1890.

Miller, Francis Trevelyan. *Thomas A. Edison, Benefactor of Mankind: The Romantic Life Story of the World's Greatest Inventor.* Philadelphia: John C. Winston Co., 1931.

Nerney, Mary Childs. *Thomas A. Edison: A Modern Olympian*. New York: H. Smith and R. Haas, 1934

Rolt-Wheeler, Francis. *Thomas Alva Edison*. New York: Macmillan Co., 1915.

Simonds, William Adams. *Edison: His Life, His Work, His Genius*. New York: Bobbs-Merrill Co., 1934.

Tate, Alfred O. *Edison's Open Door: The Life Story of Thomas A. Edison, A Great Industrialist*. New York: E. P. Dutton & Co., 1938.

3. Select Juvenile Works

Adair, Gene. *Thomas Alva Edison: Inventing the Electric Age*. Oxford: Oxford University Press, 1996.

Barnham, Kay. *Thomas Edison*. Chicago: Heinemann-Raintree, 2014.

Barretta, Gene. *Timeless Thomas: How Thomas Edison Changed Our Lives*. New York: Henry Holt, 2012.

Burgan, Michael. *Thomas Alva Edison: Great American Inventor*. Minneapolis: Compass Point Books, 2007.

Colbert, David. *Thomas Edison*. New York: Simon & Schuster, 2008.

Graham, Amy. *Thomas Edison: Wizard of Light and Sound*. Berkeley Heights, NJ: Enslow, 2006.

Jenner, Caryn. *Thomas Edison: The Great Inventor*. New York: DK Publishing, 2007.

Krieg, Katherine. *Thomas Edison: World-Changing Inventor*. Minneapolis: Core Library, 2014.

Meadowcroft, William H. *The Boy's Life of Edison*. New York: Harper & Brothers, 1911, 1921.

Mitchell, Barbara. *The Wizard of Sound: A Story about Thomas Edison*. Minneapolis: Millbrook Press, 2012.

Mortensen, Lori. *Thomas Edison: Inventor, Scientist, and Genius*. Minneapolis: Picture Window Books, 2007.

Pederson, Charles E. *Thomas Edison*. Edina, MN: ABDO Publishing, 2008.

Rausch, Monica L. *Thomas Edison and the Lightbulb*. Milwaukee, WI: Weekly Reader Early Learning Library, 2007.

Sonneborn, Liz. *The Electric Light: Thomas Edison's Illuminating Invention*. New York: Chelsea House, 2007.

Tagliaferro, Linda. *Thomas Edison: Inventor of the Age of Electricity*. Minneapolis: Lerner Publications Co., 2003.

Woodside, Martin. *Thomas Edison: The Man Who Lit Up the World*. New York: Sterling Publishing Company, 2007.

Subject Areas Related to Edison's Work and Influence

1. Electric Light and Power

Arapostathis, Stathis. "Consulting Engineers in the British Electric Light and Power Industry, ca. 1880–1914." PhD diss., University of Oxford, 2006.

———. "Dynamos, Tests, and Consulting in the Career of John Hopkinson in the 1880s." *Annales Historiques de l'Électricité* 1 (2007): 8–31.

———. "Electrical Technoscience and Physics in Transition, 1880–1920." *Studies in History and Philosophy of Science* 44 (2013): 202–11.

———. "Morality, Locality and 'Standardization' in the Work of British Consulting Electrical Engineers, 1880–1914." *History of Technology* 28 (2016): 53–74.

Arapostathis, Stathis, and Graeme Gooday. *Patently Contestable: Electrical Technologies and Inventor Identities on Trial in Britain*. Cambridge, MA: MIT Press, 2013.

Bazerman, Charles. *The Language of Edison's Light*. Cambridge, MA: MIT Press, 1999.

Bright, Arthur Aaron. *The Electric-Lamp Industry: Technological Change and Economic Development from 1800 to 1947*. New York: Macmillan Co., 1949.

Brittain, James E. "The International Diffusion of Electrical Power Technology, 1870–1920." *Journal of Economic History* 34, no. 1 (1974): 108–21.

Brox, Jane. *Brilliant: The Evolution of Artificial Light*. Boston: Houghton Mifflin Harcourt, 2010.

Carlat, Louis, and Daniel Weeks. "'New and Untried Hands': Thomas Edison's Electrification of Pennsylvania Towns, 1883–85." *Pennsylvania Magazine of History and Biography* 139 (2015): 293–321.

Carlson, W. Bernard. "Alternating Current Comes to America: Edison, Thomson, and Tesla." In *Proceedings of the International Symposium on Galileo Ferraris and the Conversion of Energy: Developments of Electrical Engineering over a Century.* Turin: Istituto Elettrotecnico Nazionale Galileo Ferraris, 2000, 143–78.

———. "Competition and Consolidation in the Electrical Manufacturing Industry." In *Technological Competitiveness: Contemporary and Historical Perspectives on the Electrical, Electronics, and Computer Industries,* ed. William Aspray, 287–311. Piscataway, NJ: IEEE Press, 1993.

Carlson, W. Bernard, and A. J. Millard. "Defining Risk within a Business Context: Thomas A. Edison, Elihu Thomson, and the A.C.–D.C. Controversy, 1885–1900." In *The Social and Cultural Construction of Risk,* ed. B. B. Johnson and V. T. Covello, 275–93. Boston: Reidel Publishing, 1987.

Cunningham, Joseph J. "STARS: Manhattan Electric Power Distribution, 1881–1901." *Proceedings of the IEEE* 103, no. 5 (2015): 850–58.

David, Paul A. "The Hero and the Herd in Technological History: Reflections on Thomas Edison and the Battle of the Systems." In *Favorites of Fortune: Technology, Growth, and Economic Development since the Industrial Revolution,* ed. Patrice Higgonet, David S. Landes, and Henry Rosovsky, 72–119. Cambridge, MA: Harvard University Press, 1998.

———. 1992. "Heroes, Herds, and Hysteresis in Technological History: Thomas Edison and 'The Battle of the Systems' Reconsidered." *Industrial and Corporate Change* 1 (1992): 129–80.

Essig, Mark. *Edison and the Electric Chair: A Story of Light and Death.* New York: Walker & Co., 2003.

Fox, Robert. "Thomas Edison's Parisian Campaign: Incandescent Lighting and the Hidden Face of Technology Transfer." *Annals of Science* 53 (1996): 157–93.

Freeberg, Ernest. *The Age of Edison: Electric Light and the Invention of Modern America.* New York: Penguin Press, 2013.

Friedel, Robert D., Paul Israel, and Bernard S. Finn. *Edison's Electric Light: The Art of Invention.* Baltimore: Johns Hopkins University Press, 2010.

Goldfarb, Brent. "Diffusion of General-Purpose Technologies: Understanding Patterns of the Electrification of U.S. Manufacturing 1880–1930." *Industrial and Corporate Change* 14, no. 5 (2005): 745–73.

Gooday, Graeme. *Domesticating Electricity: Technology, Uncertainty and Gender, 1880–1914.* London: Pickering & Chatto, 2008.

Guagnini, Anna. "A Bold Leap into Electric Light: The Creation of the Società Italiana Edison, 1880–1886." *History of Technology* 32 (2014): 155–90.

Hargadon, Andrew, and Douglas Yellowlees. "When Innovations Meet Institutions: Edison and the Design of the Electric Light." *Administrative Science Quarterly* 46, no. 3 (2001): 476–501.

Hausman, William J., Peter Hertner, and Mira Wilkins. *Global Electrification: Multinational Enterprise and International Finance in the History of Light and Power, 1878–2007.* Cambridge: Cambridge University Press, 2008.

Hausman, William J., and John L. Neufeld. "Battle of the Systems Revisited: The Role of Copper." *IEEE Technology and Society Magazine* 11, no. 3 (1992): 18–25.

Hellrigel, Mary Ann. "The Quest to Be Modern: The Adoption of Electric Light, Heat, and Power Technology in Small-Town America, 1883–1929." PhD diss., Case Western Reserve University, 1997.

Hughes, Thomas Parke. "British Electrical Industry Lag: 1882–1888." *Technology and Culture* 3, no. 1 (1962): 27–44.

———. "The Electrification of America: The System Builders." *Technology and Culture* 20 (1979): 124–61.

———. "Harold P. Brown and the Executioner's Current: An Incident in the AC-DC Controversy." *Business History Review* 32 (1958): 143–65.

———. *Networks of Power: Electrification in Western Society, 1880–1930.* Baltimore: Johns Hopkins University Press, 1983.

Hunt, Bruce J. *Pursuing Power and Light: Technology and Physics from James Watt* to *Albert Einstein*. Baltimore: Johns Hopkins University Press, 2010.

Jonnes, Jill. *Empires of Light: Edison, Tesla, Westinghouse, and the Race to Electrify the World*. New York: Random House, 2003.

Klein, Maury. *The Power Makers: Steam, Electricity, and the Men Who Invented Modern America*. New York: Bloomsbury Publishing, 2010.

Kline, Ronald, and Thomas C. Lassman. "Competing Research Traditions in American Industry: Uncertain Alliances between Engineering and Science at Westinghouse Electric, 1886–1935." *Enterprise and Society* 6 (2005): 601–45.

Marvin, Carolyn. "Dazzling the Multitude: Imagining the Electric Light as a Communications Medium." In *Imagining Tomorrow: History, Technology, and the American Future*, ed. Joseph Corn, 202–17. Cambridge, MA: MIT Press, 1986.

McDonnel, J. W. "The Electric Light and the Future: American Perceptions and Expectations, 1879–1890." MA thesis, University of Ottawa, 1997.

McGuire, Patrick. "Money and Power: Financiers and the Electric Manufacturing Industry, 1878–1896." *Social Science Quarterly* 71, no. 3 (1990): 510.

Millard, Andre. "Thomas Edison, the Battle of the Systems and the Persistence of Direct Current." *Material History Bulletin* 36 (1992): 18–28.

Moran, Richard. *Executioner's Current: Thomas Edison, George Westinghouse, and the Invention of the Electric Chair*. New York: Knopf, 2002.

Nam, Moon-Hyon. "Early History of Korean Electric Light and Power Development." In *2007 Conference on the History of Electric Power*, 192–200. Piscataway, NJ: IEEE, 2007.

Nye, David E. *American Illuminations: Urban Lighting, 1800–1920*. Cambridge, MA: MIT Press, 2019.

———. *Electrifying America: Social Meanings of a New Technology*. Cambridge, MA: MIT Press, 1991.

Otter, Chris. *The Victorian Eye: A Political History of Light and Vision in Britain, 1800–1910*. Chicago: University of Chicago Press, 2008.

Passer, Harold C. *The Electrical Manufacturers, 1875–1900: A Study in Competition, Entrepreneurship, Technical Change, and Economic Growth*. Cambridge, MA: Harvard University Press, 1953.

Platt, Harold L. *The Electric City: Energy and the Growth of the Chicago Area, 1880–1930*. Chicago: University of Chicago Press, 1991.

Reynolds, Terry, and Theodore Bernstein. "The Damnable Alternating Current." *Proceedings of the IEEE* 64 (1976): 1339–43.

———. "Edison and the 'Chair.'" *IEEE Technology and Society Magazine* 8 (1989): 19–28.

Rosenberg, Robert. "Academic Physics and the Origins of Electrical Engineering in America." PhD diss., Johns Hopkins University, 1990.

———. "Test Men, Experts, Brother Engineers, and Members of the Fraternity: Whence the Early Electrical Work Force?" *IEEE Transactions on Education* E-27, no. 4 (1984): 203–10.

Schatzberg, Eric. "The Mechanization of Urban Transit in the United States." In *Technological Competitiveness: Contemporary and Historical Perspectives on the Electrical, Electronics, and Computer Industries*, ed. William Aspray, 225–42. Piscataway, NJ: IEEE Press, 1993.

Schivelbusch, Wolfgang. *Disenchanted Night: The Industrialization of Light in the Nineteenth Century*. Berkeley: University of California Press, 1995.

Sullivan, Joseph P. "Fearing Electricity: Overhead Wire Panic in New York City." *IEEE Technology and Society Magazine* 14, no. 3 (1996): 8–16.

Usselman, Steven W. "From Novelty to Utility: George Westinghouse and the Business of Innovation during the Age of Edison." *Business History Review* 66 (1992): 251–304.

Vincenti, Walter G. "The Technical Shaping of Technology: Real-World Constraints and Technical Logic in Edison's Electrical Lighting System." *Social Studies of Science* 25, no. 3 (1995): 553–74.

2. Phonograph and Sound Recording

Barnett, Kyle. *Record Cultures: The Transformation of the U.S. Recording Industry.* Ann Arbor: University of Michigan Press, 2020.

Brooks, Tim. *Lost Sounds: Blacks and the Birth of the Recording Industry, 1890–1919.* Urbana: University of Illinois Press, 2004.

Camlot, Jason. "Early Talking Books: Spoken Recordings and Recitation Anthologies, 1880–1920." *Book History* 6 (2003): 147–73.

DeGraaf, Leonard. "Confronting the Mass Market: Thomas Edison and the Entertainment Phonograph." *Business and Economic History* 24 (1995): 88–96.

De Lautour, Reuben. "Sound, Reproduction, Mysticism: Thomas Edison and the Mythology of the Phonograph." *Música em Contexto* 1 (2015): 23–53.

Feaster, Patrick. "The Artifice of Nineteenth-Century Phonographic Business Dictation." *The Velvet Light Trap* 72 (2010): 3–16.

———. "'The Following Record': Making Sense of Phonographic Performance, 1877–1908." PhD diss., Indiana University, 2007.

———. "Phonographic Etiquette; or 'The Spirit First Moves Mister Knowles.'" *Victorian Review* 38 (2012): 18–23.

———. "'Rise and Obey the Command': Performative Fidelity and the Exercise of Phonographic Power." *Journal of Popular Music Studies* 24 (2012): 357–95.

———. "'Things Enough for So Many Dolls to Say': A Cultural History of the Edison Talking Doll Record." Online publication hosted by U.S. National Park Service (2015; https://www.nps.gov/edis/learn/photosmultimedia/a-cultural-history-of-the-edison-talking-doll-record.htm).

Frow, George L. *Edison Cylinder Phonograph Companion.* Woodland Hills, CA: Stationary X-Press, 1994.

———. *Edison Disc Phonographs and the Diamond Discs.* Los Angeles: Mulholland Press, 2001.

Gitelman, Lisa. "The Phonograph's New Media Publics." In *The Sound Studies Reader,* ed. Jonathan Sterne, 283–303. New York: Routledge, 2012.

———. *Scripts, Grooves, and Writing Machines: Representing Technology in the Edison Era.* Stanford, CA: Stanford University Press, 1999.

Grack, Tim, with Frank Hoffman. *Popular American Recording Pioneers, 1895–1925.* New York: Haworth Press, 2000.

Hochman, Brian. *Savage Preservation: The Ethnographic Origins of Modern Media Technology.* Minneapolis: University of Minnesota Press, 2014.

Hui, Alexandra. "Lost: Thomas Edison's Mood Music Found: New Ways of Listening." *Endeavor* 38 (2014): 139–42.

Israel, Paul. "The Unknown History of the Tinfoil Phonograph." *NARAS Journal* 8 (1997–1998): 29–42.

Kelleher, Kevin Daniel. "The Contributions of Thomas Alva Edison to Music Education," PhD diss., Old Dominion University, 2013.

Kenney, William Howland. *Recorded Music in American Life: The Phonograph and Popular Memory, 1890–1945.* Oxford: Oxford University Press, 1999.

Koenigsberg, Allen. *Edison Cylinder Records, 1889–1912, with an Illustrated History of the Phonograph.* New York: Stellar Productions, 1969.

———. *The Patent History of the Phonograph, 1877–1912.* Brooklyn, NY: APM Press, 1990, 1992.

Korzi, Michael J. "William Howard Taft, the 1908 Election, and the Future of the American Presidency." *Congress & the Presidency* 43 (2016): 227–54.

Leppert, Richard. "Caruso, Phonography, and Operatic Fidelities: Regimes of Musical Listening, 1904–1929." In *Aesthetic Technologies of Modernity, Subjectivity, and Nature: Opera, Orchestra, Phonograph, Film,* chap. 3. Berkeley: University of California Press, 2015.

Magoun, Alexander. "Shaping the Sound of Music: The Evolution of the Phonograph Record, 1877–1950." PhD diss., University of Maryland, 2000.

Martland, Peter. *Recording History: The British Record Industry, 1888–1931*. Lanham, MD: Scarecrow Press, 2013.

Millard, Andre. *America On Record: A History of Recorded Sound*. Cambridge: Cambridge University Press, 1995.

Morton, David. *Sound Recording: The Life Story of a Technology*. Baltimore: Johns Hopkins University Press, 2006.

Picker, John. "The Victorian Aura of the Recorded Voice." *New Literary History* 32 (2001): 770.

———. *Victorian Soundscapes*. Oxford and New York: Oxford University Press, 2003.

Radick, Gregory. "Morgan's Canon, Garner's Phonograph, and the Evolutionary Origins of Language and Reason." *British Journal for the History of Science* 33 (2000): 3–23.

Read, Oliver, and Walter L. Welch. *From Tin Foil to Stereo: Evolution of the Phonograph*. Indianapolis: H. W. Sams, 1959, 1976.

Rubery, Matthew. "Canned Literature: The Book after Edison." *Book History* 16 (2013): 215–45.

Selfridge-Field, Eleanor. "Experiments with Melody and Meter, or the Effects of Music: The Edison-Bingham Music Research." *The Musical Quarterly* 81 (1997): 291–310.

Siefert, Marsha. "Aesthetics, Technology, and the Capitalization of Culture: How the Talking Machine Became a Musical Instrument." *Science in Context* 8 (1995): 417–49.

Thompson, Emily. "Is It Real or Is It a Machine?" *American Heritage of Invention & Technology* 12 (1997): 50–56.

———. "Machines, Music, and the Quest for Fidelity: Marketing the Edison Phonograph in America, 1877–1925." *The Musical Quarterly* 79 (1995): 131–71.

Vest, Jacques. "Vox Machinae: Phonographs and the Birth of Sonic Modernity, 1870–1930." PhD diss., University of Michigan, 2018.

Welch, Walter L., and Leah Brodbeck Stenzel Burt. *From Tinfoil to Stereo: The Acoustic Years of the Recording Industry, 1877–1929*. Gainesville: University Press of Florida, 1994.

Wile, Raymond R. "The Automatic Phonograph Exhibition Company and the Beginnings of the Nickel-in-the-Slot Phonograph." *Association for Recorded Sound Collections Journal* 33 (Spring 2002): 1–20.

———. "Jack Fell Down and Broke His Crown: The Fate of the Edison Phonograph Toy Manufacturing Company." *Association for Recorded Sound Collections Journal* 19, nos. 2–3 (1987): 5–36.

———. "The Metropolitan Phonograph Company." *Association for Recorded Sound Collections Journal* 34 (Spring 2003): 1–13.

3. Motion Pictures

*Note that an extensive article literature, not listed below, appears in *Cinema Journal* and *Film History*.

Abel, Richard. *The Red Rooster Scare: Making Cinema American, 1900–1910*. Berkeley: University of California Press, 1999.

———, ed. *Silent Film*. New Brunswick, NJ: Rutgers University Press, 1996.

Abel, Richard, and Rick Altman, eds. *The Sounds of Early Cinema*. Bloomington: Indiana University Press, 2001.

Altman, Rick. *Silent Film Sound*. New York: Columbia University Press, 2004.

Anderson, Robert Jack. "The Motion Picture Patents Company." PhD diss., University of Wisconsin, Madison, 1983.

Auerbach, Jonathan. *Body Shots: Early Cinema's Incarnations*. Berkeley: University of California Press, 2007.

Balio, Tino, ed. *The American Film Industry*. Madison: University of Wisconsin Press, 1985.

Bowser, Eileen. *The Transformation of Cinema, 1907–1915*. New York: Scribner, 1990.

Braun, Marta, Charlie Keil, Rob King, Paul Moore, and Louis Pelletier, eds. *Beyond the Screen: Institutions, Networks and Publics of Early Cinema*. New Barnet: John Libbey, 2012.

Brown, Richard. "A New Look at Old History—The Kinetoscope: Fraud and Market Development in Britain in 1895." *Early Popular Visual Culture* 10 (2012): 407–39.

Brown, Richard, and Barry Anthony. *The Kinetoscope: A British History*. Bloomington: Indiana University Press, 2017.

Brownlow, Kevin. *Behind the Mask of Innocence: Sex, Violence, Prejudice, Crime: Films of Social Conscience in the Silent Era*. Berkeley: University of California Press, 1990.

Carlson, W. Bernard. "Artifacts and Frames of Meaning: Thomas A. Edison, His Managers, and the Cultural Construction of Motion Pictures." In *Shaping Technology/Building Society: Studies in Sociotechnical Change*, ed. Wiebe E. Bijker and John Law, 175–98. Cambridge, MA: MIT Press, 1992.

Carlson, W. Bernard, and Michael E. Gorman. "Understanding Invention as a Cognitive Process: The Case of Thomas Edison and Early Motion Pictures, 1888–91." *Social Studies of Science* 20 (1990): 387–430.

Christie, Ian. *Robert Paul and the Origins of British Cinema*. Chicago: University of Chicago Press, 2019.

Coffman, Elizabeth. "Women in Motion: Loie Fuller and the 'Interpenetration' of Art and Science." *Camera Obscura* 17 (2002): 1–104.

Decherney, Peter. *Hollywood's Copyright Wars: From Edison to the Internet*. New York: Columbia University Press, 2012.

Fischer, Paul. *The Man Who Invented Motion Pictures: A True Tale of Obsession, Murder, and the Movies*. New York: Simon & Schuster, 2022.

Frykholm, Joel. *George Kleine and American Cinema: The Movie Business and Film Culture in the Silent Era*. London: British Film Institute, 2019.

Gaines, Jane M. *Pink-Slipped: What Happened to Women in the Silent Film Industries?* Urbana: University of Illinois Press, 2018.

Gaudio, Michael. "Dancing for the Kinetograph: The Lakota Ghost Dance and the Silence of Early Cinema." In *Sound, Image, Silence: Art and the Aural Imagination in the Atlantic World*, chap. 5. Minneapolis: University of Minnesota Press, 2019.

Gaudreault, André, ed. *American Cinema 1890–1909: Themes and Variations*. New Brunswick, NJ: Rutgers University Press, 2009.

Gaudreault, André, Nicolas Dulac, and Santiago Hidalgo, eds. *A Companion to Early Cinema*. New York: John Wiley & Sons, 2012.

Grieveson, Lee, and Peter Kramer. *The Silent Cinema Reader*. London: Routledge, 2004.

Hendricks, Gordon. *The Edison Motion Picture Myth*. Berkeley: University of California Press, 1961.

Jacobson, Brian R. *Studios Before the System: Architecture, Technology, and the Emergence of Cinematic Space*. New York: Columbia University Press, 2015.

Keil, Charlie. *Early American Cinema in Transition: Story, Style, and Filmmaking, 1907–1913*. Madison: University of Wisconsin Press, 2001.

Keil, Charlie, and Shelley Stamp, eds. *American Cinema's Transitional Era: Audiences, Institutions, Practices*. Berkeley: University of California Press, 2004.

Kessler, Frank, and Nanna Verhoeff. *Networks of Entertainment: Early Film Distribution 1895–1915*. New Barnet: John Libbey, 2007.

Lipton, Lenny. *The Cinema in Flux: The Evolution of Motion Picture Technology from the Magic Lantern to the Digital Era*. New York: Springer, 2020.

Mathews, Nancy Mowll, Charles Musser, and Marta Braun. *Moving Pictures: American Art and Early Film, 1880–1910*. Manchester, VT: Hudson Hills Press, 2005.

Musser, Charles. *Before the Nickelodeon: Edwin S. Porter and the Edison Manufacturing Company*. Berkeley: University of California Press, 1991.

———. *Edison Motion Pictures, 1890–1900: An Annotated Filmography*. Washington, DC: Smithsonian Institution Press, 1997.

———. *The Emergence of Cinema: The American Screen to 1907*. Berkeley: University of California Press, 1990.

———. *High-Class Moving Pictures: Lyman H. Howe and the Forgotten Era of Traveling Exhibition, 1880–1920*. Princeton, NJ: Princeton University Press, 1991.

———. *Thomas A. Edison and His Kinetographic Motion Pictures*. New Brunswick,

NJ: Rutgers University Press for the Friends of Edison National Historic Site, 1995.

Orgeron, Devin, Marsha Orgeron, and Dan Streible. *Learning with the Lights Off: Educational Film in the United States*. Oxford: Oxford University Press, 2012.

Phillips, Ray. *Edison's Kinetoscope and Its Films: A History to 1896*. Westport, CT: Greenwood Press, 1997.

Ramsaye, Terry. *A Million and One Nights: A History of the Motion Picture*. New York: Simon & Schuster, 1954.

Robinson, David. *From Peep Show to Palace: The Birth of American Film*. New York: Columbia University Press, 1996.

Spehr, Paul C. *The Man Who Made Movies: W. K. L. Dickson*. New Barnet Herts., U.K.: John Libbey, 2008.

Strauven, Wanda, ed. *The Cinema of Attractions Reloaded*. Amsterdam: Amsterdam University Press, 2006.

Streible, Dan *Fight Pictures: A History of Boxing and Early Cinema*. Berkeley: University of California Press, 2008.

4. Telecommunications

Beauchamp, Christopher. *Invented by Law: Alexander Graham Bell and the Patent That Changed America*. Cambridge, MA: Harvard University Press, 2015.

Butrica, Andrew J. "From Inspecteur to Ingenieur: Telegraphy and the Genesis of Electrical Engineering in France, 1845–1881." PhD diss., Iowa State University, 1986.

Carlson, W. Bernard. "Invention and Evolution: The Case of Edison's Sketches of the Telephone." In *Technological Innovation as an Evolutionary Process*, ed. J. Ziman, 137–58. New York: Cambridge University Press, 2000.

Carlson, W. Bernard, and Michael E. Gorman, "A Cognitive Framework to Understand Technological Creativity: Bell, Edison, and the Telephone." In *Inventive Minds: Creativity in Technology*, ed. Robert J. Weber, 48–79. Oxford: Oxford University Press, 1992.

———. "Thinking and Doing at Menlo Park: Edison's Development of the Telephone, 1876–1878." In *Working at Inventing: Thomas Edison and the Menlo Park Experience*, ed. William S. Pretzer, 84–99. Dearborn, MI: Henry Ford Museum & Greenfield Village, 1989; repr., Baltimore: Johns Hopkins University Press, 2002.

DuBoff, Richard B. "The Telegraph and the Structure of Markets in the United States, 1845–1890." In *Research in Economic History*, vol. 8, ed. Paul Uselding, 253–77. Greenwich, CT: JAI Press, 1983.

Fagen, M. D., ed. *A History of Engineering and Science in the Bell System: The Early Years (1875–1925)*. N.p.: Bell Telephone Laboratories, Inc., 1975.

Gabler, Edwin. *The American Telegrapher: A Social History, 1860–1900*. New Brunswick, NJ: Rutgers University Press, 1988.

Gorman, Michael E., Matthew M. Mehalik, W. Bernard Carlson, and Michael Oblon. "Alexander Graham Bell, Elisha Gray and the Speaking Telegraph: A Cognitive Comparison." *History of Technology* 15 (1993): 1–56.

Hochfelder, David. *The Telegraph in America, 1832–1920*. Baltimore: Johns Hopkins University Press, 2012.

Hounshell, David. "Elisha Gray and the Telephone: On the Disadvantages of Being an Expert." *Technology and Culture* 16 (1975): 133–61.

Israel, Paul. *From Machine Shop to Industrial Laboratory: Telegraphy and the Changing Context of American Invention, 1830–1920*. Baltimore: Johns Hopkins University Press, 1992.

John, Richard R. *Network Nation: Inventing American Telecommunications*. Cambridge, MA: Harvard University Press, 2010.

Kieve, Jeffrey L. *The Electric Telegraph: A Social and Economic History*. Newton Abbot: David & Charles, 1973.

King, W. James. "The Development of Electrical Technology in the Nineteenth Century: 2. The Telegraph and the Telephone." *United States Museum Bulletin*, no. 228. Washington, DC: Smithsonian Institution, 1962.

Nye, David E. "Shaping Communication Networks: Telegraph, Telephone, Computer." *Social Research* 64 (1997): 1067–92.

Smith, George David. *The Anatomy of a Business Strategy: Bell, Western Electric, and the Origins of the American Telephone Industry*. Baltimore: Johns Hopkins University Press, 1985.

Tosiello, Rosario J. *The Birth and Early Years of the Bell Telephone System, 1876–1880*. New York: Arno Press, 1979. (Orig. pub. 1971.)

Willis, Ian. "Instrumentalizing Failure: Edison's Invention of the Carbon Microphone." *Annals of Science* 64 (2007): 383–409.

Wolff, Joshua D. *Western Union and the Creation of the American Corporate Order, 1845–1893*. Cambridge: Cambridge University Press, 2013.

5. Batteries and Electric Automobiles

Black, Edwin. *Internal Combustion: How Corporations and Governments Addicted the World to Oil and Derailed the Alternatives*. New York: St. Martin's Press, 2006.

Kirsch, David A. *The Electric Vehicle and the Burden of History*. New Brunswick, NJ: Rutgers University Press, 2000.

Mom, Gijs. *The Electric Vehicle: Technology and Expectations in the Automobile Age*. Baltimore: Johns Hopkins University Press, 2004.

Schallenberg, Richard H. *The Storage Battery: Bottled Energy. Electrical Engineering and the Evolution of Chemical Energy Storage*. Philadelphia: American Philosophical Society, 1982.

Segrave, Kerry. *The Electric Automobile in America, 1890–1922: A Social History*. Jefferson, NC: McFarland, 2019.

6. Materials Processing (Ore Milling, Cement, and Rubber)

Bray, Matt. "A Company and a Community: The Canadian Copper Company and Sudbury, 1886–1902. In *At the End of the Shift: Mines and Single-Industry Towns in Northern Ontario*, ed. Matt Bray and Ashley Thomson, 23–44. Toronto: Dundurn Press, 1992.

Carlson, W. Bernard. "Edison in the Mountains: The Magnetic Ore Separation Venture, 1879–1900." *History of Technology* 8 (1983): 37–59.

Dublin, Thomas, and Walter Licht. *The Face of Decline: The Pennsylvania Anthracite Region in the Twentieth Century*. Ithaca, NY: Cornell University Press, 2005.

Finlay, Mark R. *Growing American Rubber: Strategic Plants and the Politics of National Security*. New Brunswick, NJ: Rutgers University Press, 2009.

Jandl, H. Ward, John A. Burns, and Michael J. Auer, *Yesterday's Houses of Tomorrow: Innovative American Homes 1850 to 1950*. Washington, DC: Preservation Press, 1991.

Johnson, Rodney P. *Thomas Edison's "Ogden Baby": The New Jersey and Pennsylvania Concentrating Works*. Highland Lakes, NJ: privately printed, 2004.

Leidy, Thomas, and Donald R. Shenton. "Titan in Berks: Edison's Experiments in Iron Concentration." *Historical Review of Berks County* 13 (1958): 104–10.

Misa, Thomas J. *A Nation of Steel: The Making of Modern America, 1865–1925*. Baltimore: Johns Hopkins University Press, 1995.

Newell, Dianne. *Technology on the Frontier: Mining in Old Ontario*. Vancouver: University of British Columbia Press, 2014.

Peterson, Michael. "Thomas Edison, Failure." *American Heritage of Invention & Technology* 6 (1991): 8–14.

———. "Thomas Edison's Concrete House." *American Heritage of Invention & Technology* 11 (1996): 50–56.

Weeks, Daniel. "'Practically Intoxicated': Thomas Edison and the Bechtelsville Ore-Milling Experiment." *Pennsylvania Magazine of History and Biography* 145 (2021): 66–87.

7. World War I

Franklin, H. Bruce. *War Stars: The Super Weapon and the American Imagination*. Oxford: Oxford University Press, 1988.

Jeffrey, Thomas E. "'Commodore' Edison Joins the Navy: Thomas Alva Edison and the Naval Consulting Board." *Journal of Military History* 80 (2016): 411–45.

————. *From Phonographs to U-Boats: Edison and His "Insomnia Squad" in Peace and War, 1911–1919*. Bethesda, MD: Lexis-Nexis, 2008.

————. "'When the Cat Is Away the Mice Will Work': Thomas Alva Edison and the Insomnia Squad." *New Jersey History* 125, no. 2 (2010): 1–22.

McBride, William M. "The 'Greatest Patron of Science'?: The Navy-Academia Alliance and U.S. Naval Research, 1896–1923." *Journal of Military History* 56 (1992): 7–33.

Scott, Lloyd. N. *Naval Consulting Board of the United States*. Washington, DC: Government Printing Office, 1920.

Thelander, Theodore A. "Josephus Daniels and the Publicity Campaign for Naval and Industrial Preparedness before World War I." *North Carolina Historical Review* 43 (1996): 316–32.

Van Keuren, David K. "Advancing Scholarship in Wartime: The World War I Research Experience and Its Impact on American Higher Education, 1900–1925." PhD diss, University of Illinois Urbana-Champaign, 1997.

————. "Science, Progressivism, and Military Preparedness: The Case of the Naval Research Laboratory, 1915–1923." *Technology and Culture* 33 (1992): 710–36.

Worthington, Daniel E. "Inventive Genius or Scientific Research?" *Swords & Plowshares* 6 (1991): 11–14.

8. Miscellaneous Topics

Baron, David. *American Eclipse: A Nation's Epic Race to Catch the Shadow of the Moon and Win the Glory of the World*. New York: Norton, 2017.

Boxman, Raymond L. "Early History of Vacuum Arc Deposition." *IEEE Transactions on Plasma Science* 29 (2001): 759–61.

Cooper, Jill E. "Intermediaries and Invention: Business Agents and the Edison Electric Pen and Duplicating Press." *Business and Economic History* 25 (1996): 30–142.

Dennis, Paul M. "The Edison Questionnaire." *Journal of the History of the Behavioral Sciences* 20 (1984): 23–37.

Genalo, Lawrence J., Denise A. Schmidt, and Melanie Schiltz. "Piaget and Engineering Education." *Proceedings of the ASEE Annual Conference*, June 2004.

Hartmann, Thom. *The Edison Gene: ADHD and the Gift of the Hunter Child*. Rochester, VT: Park Street Press, 2003.

Herzig, Rebecca. "In the Name of Science: Suffering, Sacrifice, and the Formation of American Roentgenology." *American Quarterly* 53 (2001): 563–89.

Hill, George J. *Edison's Environment: Invention and Pollution in the Career of Thomas Edison*. Morristown, NJ: New Jersey Heritage Press, 2007.

Limb, Charles J., and Lawrence R. Lustig. "Hearing Loss and the Invention of the Phonograph: The Story of Thomas Alva Edison." *Otology & Neurotology* 23, no. 1 (2002): 96–101.

Miller, Stuart H. "A Medical Investigation of Edison's Deafness." *Sound Box*, June 2004, 16–20.

Pretzer, William S., George E. Rogers, and Jeffery Bush. "A Model Technology Educator: Thomas A. Edison." *Technology Teacher*, September 2007, 27–31.

Samuels, David W. "Edison's Ghost." *Music & Politics* 10 (2016): 1–8.

Teel, Scott, and Vincent Monastra. *Defending and Parenting Children Who Learn Differently: Lessons from Edison's Mother*. Westport, CT: Praeger, 2007.

Tselos, George D. "New Jersey's Thomas Edison and the Fluoroscope." *New Jersey Medicine* 92 (1995): 731–33.

Unwin, Peter. "'An Extremely Useful Invention': Edison's Electric Pen and the Unravelling of Old and New Media." *Convergence* 25 (2019): 607–26.

Waits, Robert K. "Edison's Vacuum Coating Patents." *Journal of Vacuum Science & Technology A: Vacuum, Surfaces, and Films* 19 (2001): 1666–73.

9. Laboratories, Patents, and Research Methods

Buonanno, Joseph F. "Thomas A. Edison: Wizard of Menlo Park, or Ordinary Thinker? A ity." PhD diss., Temple University, 2005.

Carlson, W. Bernard. "Building Thomas Edison's Laboratory at West Orange, New Jersey: A Case Study in Using Craft Knowledge for Technological Invention, 1886–1888." *History of Technology* 13 (1991): 150–67.

Hounshell, David A. "Edison and the Pure Science Ideal in 19th-Century America." *Science* 207 (1980): 612–17.

Hughes, Thomas Parke. *American Genesis: A Century of Invention and Technological Enthusiasm.* Chicago: University of Chicago Press, 2004.

Israel, Paul B. "Claim the Earth: Protecting Edison's Inventions at Home and Abroad." In *Knowledge Management and Intellectual Property: Concepts, Actors and Practices from the Past to the Present*, ed. Graham Dutfield and Stathis Arapostathis, 19–43. Cheltenham: Edward Elgar, 2013.

———. "Inventing Industrial Research: Thomas Edison and the Menlo Park Laboratory." *Endeavor* 26 (2002): 48–54.

Israel, Paul B., and Robert Rosenberg. "Patent Office Records as a Historical Source: The Case of Thomas Edison." *Technology and Culture* 32 (1991): 1094–1101.

Jeffrey, Thomas E. "'When the Cat Is Away the Mice Will Work': Thomas Alva Edison and the Insomnia Squad." *New Jersey History* 125 (2010): 1–21.

Jehl, Francis. *Menlo Park Reminiscences.* 3 vols. Dearborn, MI: Edison Institute, 1937–1941.

Kline, Ronald, and Thomas C. Lassman. "Competing Research Traditions in American Industry: Uncertain Alliances between Engineering and Science at Westinghouse Electric, 1886–1935." *Enterprise and Society* 6 (2005): 601–45.

Millard, Andre J. *Edison and the Business of Innovation.* Baltimore: Johns Hopkins University Press, 1990.

Millard, Andre J., Duncan Hay, and Mary Grassick. *Historic Furnishings Report: Edison Laboratory.* Harpers Ferry, WV: National Park Service, 1995.

Pretzer, William S., ed. *Working at Inventing: Thomas Edison and the Menlo Park Experience.* Dearborn, MI: Henry Ford Museum & Greenfield Village, 1989; repr., Baltimore: Johns Hopkins University Press, 2002.

Reich, Leonard S. "Edison, Coolidge, and Langmuir: Evolving Approaches to American Industrial Research." *Journal of Economic History* 47 (1987): 341–51.

———. *The Making of American Industrial Research: Science and Business at GE and Bell, 1876–1926.* Cambridge: Cambridge University Press, 1985.

Simonton, Dean Keith. "Thomas Edison's Creative Career: The Multilayered Trajectory of Trials, Errors, Failures, and Triumphs." *Psychology of Aesthetics, Creativity, and the Arts* 9 (2015): 2–14.

Wills, Ian. "Edison and Science: A Curious Result." *Studies in History and Philosophy of Science Part A* 40 (2009): 157–66.

———. *Thomas Edison: Success and Innovation through Failure.* Cham, Switzerland: Springer, 2019.

Weisberg, Robert W. *Creativity: Understanding Innovation in Problem Solving, Science, Invention, and the Arts.* New York: John Wiley & Sons, 2006.

Ziman, John M. *Technological Innovation as an Evolutionary Process.* Cambridge: Cambridge University Press, 2003.

10. Entrepreneurship and Economic Ideas

Axelrod, Alan. *Edison on Innovation: 102 Lessons in Creativity for Business and Beyond.* San Francisco: Jossey-Bass, 2008.

Caldicott, Sarah Miller. *Inventing the Future: What Would Thomas Edison Be Doing Today?* New York: John Wiley & Sons, 2012.

———. *Midnight Lunch: The 4 Phases of Team Collaboration Success from Thomas Edison's Lab.* New York: John Wiley & Sons, 2013.

Davis, Julie L., and Suzanne S. Harrison. *Edison in the Boardroom: How Leading Companies Realize Value from Their Intellectual Assets.* New York: John Wiley & Sons, 2001.

Evans, Harold, Gail Buckland, and David Lefer. *They Made America: From the Steam*

Engine to the Search Engine: Two Centuries of Innovators. Boston: Little, Brown, 2004.

Gelb, Michael, and Sarah Miller Caldicott. *Innovate Like Edison: The Success System of America's Greatest Inventor.* New York: Dutton, 2007.

Hammes, David. *Harvesting Gold: Thomas Edison's Experiment to Re-Invent American Money.* Silver City, NM: Richard Mahler Publications, 2012.

Hammes, David, and David T. Wills. "Edison's Plan to Reinvent Money." *Financial History Magazine* 80 (2004): 16–19.

———. "Thomas Edison's Monetary Option." *Journal of the History of Economic Thought* 28 (2006): 295–308.

Hargadon, Andrew. *How Breakthroughs Happen: The Surprising Truth about How Companies Innovate.* Boston: Harvard Business School Press, 2003.

McCormick, Blaine. *At Work with Thomas Edison: 10 Business Lessons from America's Greatest Innovator.* Irvine, CA: Entrepreneur Press, 2001.

McCormick, Blaine, and Paul Israel, "Underrated Entrepreneur: Thomas Edison's Overlooked Business Story." *IEEE Power & Energy Magazine* 3 (2005): 76–79.

Rubin, Israel. "Thomas Alva Edison's 'Treatise on National Economic Policy and Business.'" *Business History Review* 59 (1985): 433–64.

Usselman, Steven W. "From Novelty to Utility: George Westinghouse and the Business of Innovation During the Age of Edison." *Business History Review* 66 (1992): 251–304.

11. Edison Myth and Reputation

Albion, Michele Wehrwein. *The Quotable Edison.* Gainesville: University Press of Florida, 2011.

Castravelli, Lianne C. "Building beyond Limits: Fantastic Collisions between Bodies and Machines in French and English Fin-de-Siècle Literature." PhD diss., University of Montreal, 2013.

Cloonan, William. "The Splendor and Misery of the American Scientist: L'ève Future." In *Frères Ennemis: The French in American Literature, Americans in French Literature,* chap. 2. Liverpool: Liverpool University Press, 2018.

Dadley, Portia. "The Garden of Edison: Invention and the American Imagination." In *Cultural Babbage: Technology, Time and Invention,* ed. Francis Spufford and Jenny Uglow, 81–98. London: Faber & Faber, 1996.

Daly, Michael. *Topsy: The Startling Story of the Crooked-Tailed Elephant, P. T. Barnum, and the American Wizard, Thomas Edison.* New York: Atlantic Monthly Press, 2013.

Derickson, Alan. *Dangerously Sleepy: Overworked Americans and the Cult of Manly Wakefulness.* Philadelphia: University of Pennsylvania Press, 2013.

Friedel, Robert. "Perspiration in Perspective: Changing Perceptions of Genius and Expertise in American Invention." In *Inventive Minds: Creativity in Technology,* ed. Robert J. Weber and David N. Perkins, 11–26. Oxford: Oxford University Press, 1992.

———. "Serendipity Is No Accident." *Kenyon Review* 23 (2001): 36–47.

Guinn, Jeff. *The Vagabonds: The Story of Henry Ford and Thomas Edison's Ten-Year Road Trip.* New York: Simon & Schuster, 2020.

Hager, Thomas. *Electric City: The Lost History of Ford and Edison's American Utopia.* New York: Abrams Press. 2021.

Jonnes, Jill. *Eiffel's Tower: And the World's Fair Where Buffalo Bill Beguiled Paris, the Artists Quarreled, and Thomas Edison Became a Count.* New York: Viking, 2009.

Larson, Victoria Tietze. "Across Oceans, Mountains, and Spaces: Sarah Bernhardt's American Tour." *Studies in Travel Writing* 14 (2010): 159–78.

McCormick, Blaine. *Innumerable Machines in My Mind: Found Poetry in the Papers of Thomas A. Edison.* Westbrook, ME: Moon Pie Press, 2005.

Mollmann, Steven. "The War of the Worlds in the Boston Post and the Rise of American Imperialism: 'Let Mars Fire.'" *English Literature in Transition, 1880–1920* 53 (2010): 387–412.

Nadis, Fred. *Wonder Shows: Performing Science, Magic, and Religion in America*. New Brunswick, NJ: Rutgers University Press, 2005.

Nye, David. *The Invented Self: An Antibiography, from Documents of Thomas A. Edison*. Odense, Denmark: Odense University Press, 1983.

Oancea, Ana. "Edison's Modern Legend in Villiers' *L'eve Future*." *Nordlit* 15 (2012): 173–87.

Pretzer, William. "Edison's Last Breath." *Technology and Culture* 45 (2004): 679–82.

Shigeo, Sugiyama. "Biographies of Scientists and Public Understanding of Science." *AI & Society* 13 (1999): 124–34.

Simon, Linda. *Dark Light: Electricity and Anxiety from the Telegraph to the X-Ray*. Boston: Houghton Mifflin Harcourt, 2005.

Sloat, Warren. *1929: America before the Crash*. New York: Macmillan, 1979.

Villiers de l'Isle-Adam, Auguste. *Tomorrow's Eve*. Translated by Robert M. Adams. Urbana: University of Illinois Press, 1982.

Wachhorst, Wyn. *Thomas Alva Edison: An American Myth*. Cambridge, MA: MIT Press, 1981.

Weeks, Daniel. *Nearer Home: Short Histories, 1987–2019*. Eatontown, NJ: Coleridge Institute Press, 2020.

Willis, Martin T. "Edison as Time Traveler: H. G. Wells's Inspiration for His First Scientific Character." *Science Fiction Studies* 26 (1999): 284–94.

Wosk, Julie. "The "Electric Eve": Galvanizing Women in Nineteenth- and Twentieth-Century Art and Technology." *Research in Philosophy and Technology* 13 (1993): 45–56.

12. Edison Family and Fort Myers

Albion, Michele Wehrwein. *The Florida Life of Thomas Edison*. Gainesville: University Press of Florida, 2008.

Hendrick, Ellwood. *Lewis Miller: A Biographical Essay*. New York: G. P. Putnam's Sons, 1925.

Israel, Paul. "An Inventor's Wife, Mina Edison." *Timeline* 18 (May–June 2001): 2–19.

Newton, James D. *Uncommon Friends: Life with Thomas Edison, Henry Ford, Harvey Firestone, Alexis Carrel & Charles Lindbergh*. San Diego, CA: Harcourt Brace Jovanovich, 1987.

Rimer, Alexandra. *Seduced by the Light: The Mina Miller Edison Story*. Essex, CT: Lyons Press, 2023.

Smoot, Tom. *The Edisons of Fort Myers: Discoveries of the Heart*. Sarasota, FL: Pineapple Press, 2004.

Solomon, Irvin D. *Thomas Edison: The Fort Myers Connection*. Charleston, SC: Arcadia Publishing, 2001.

Thulesius, Olav. *Edison in Florida: The Green Laboratory*. Gainesville: University Press of Florida, 1997.

Venable, John D. *Out of the Shadow: The Story of Charles Edison*. East Orange, NJ: Charles Edison Fund, 1978.

Yentsch, Anne E. "Mina Miller Edison, Education, Social Reform and the Permeable Boundaries of Domestic Space, 1886–1940." In *Historical and Archaeological Perspectives on Gender Transformation: From Private to Public*, ed. Suzanne M. Spencer-Wood, 231–74. New York: Springer, 2013.

13. Edison Associates and Rivals

Acheson, Edward Goodrich. *A Pathfinder: Inventor, Scientist, Industrialist*. New York: The Press Scrap Book, 1910.

Baker, Edward Cecil. *Sir William Preece, F.R.S.: Victorian Engineer Extraordinary*. London: Hutchinson, 1976.

Baldwin, Neil. *Henry Ford and the Jews: The Mass Production of Hate*. New York: PublicAffairs, 2001.

Brinkley, Douglas G. *Wheels for the World: Henry Ford, His Company, and a Century of Progress*. New York: Viking, 2003.

Bruce, Robert V. *Bell: Alexander Graham Bell and the Conquest of Solitude*. Boston: Little, Brown, 1973.

Bryan, Ford R. *Beyond the Model T: The Other Ventures of Henry Ford*. Detroit: Wayne State University Press, 1997.

————. *Friends, Families, & Forays: Scenes from the Life and Times of Henry Ford*. Detroit: Wayne State University Press, 2002.

Buss, Dietrich G. *Henry Villard: A Study of Transatlantic Investments and Interests, 1870–1895*. New York: Arno Press, 1978.

Cahan, David. "Helmholtz and the Shaping of the American Physics Elite in the Gilded Age." *Historical Studies in the Physical and Biological Sciences* 35 (2004): 1–34.

Carlson, W. Bernard. *Innovation as a Social Process: Elihu Thomson and the Rise of General Electric, 1870–1900*. New York: Cambridge University Press, 1991.

————. "Inventor of Dreams." *Scientific American* 292 (March 2005): 78–85.

————. *Tesla: Inventor of the Electrical Age*. Princeton, NJ: Princeton University Press, 2013.

————. "Tesla, Motors, and Myths." *Phlogiston: Journal of the History of Science* 5 (2000): 77–102.

Carosso, Vincent P., and Rose C. Carosso. *The Morgans: Private International Bankers, 1854–1913*. Cambridge, MA: Harvard University Press, 1987.

Cheney, Margaret. *Tesla, Man Out of Time*. Englewood Cliffs, NJ: Prentice Hall, 1981.

Chirnside, R. C. "Sir Joseph Swan and the Invention of the Electric Lamp." *Electronics & Power* 25, no. 2 (1979): 96–100.

Cullen, John, and Alexandra Villard De Borchgrave. *Villard: The Life and Times of an American Titan*. New York: Nan A. Talese, 2001.

Dalzell, Frederick. *Engineering Invention: Frank J. Sprague and the U.S. Electrical Industry, 1880–1900*. Cambridge, MA: MIT Press, 2009.

Dood, Kendall, J. Leland I. Anderson, and Ronald R. Kline. "Tesla and the Induction Motor." *Technology and Culture* 30, no. 4 (1989): 1013–23.

Feldenkirchen, Wilfried. *Werner von Siemens: Inventor and International Entrepreneur*. Columbus: Ohio State University Press, 1994.

Fessenden, Helen May Trott. *Fessenden, Builder of Tomorrows*. New York: Coward-McCann, 1940.

Fessenden, Reginald A. 1925. "The Inventions of Reginald A. Fessenden." *Radio News* (January–November).

Ford, Henry, and Samuel Crowther. *My Life and Work*. Garden City, NY: Doubleday, Page & Co., 1922.

Fouché, Rayvon. *Black Inventors in the Age of Segregation: Granville T. Woods, Lewis H. Latimer, & Shelby J. Davidson*. Baltimore: Johns Hopkins University Press, 2003.

Gray, Charlotte. *Reluctant Genius: Alexander Graham Bell and the Passion for Invention*. New York: Arcade Publishing, 2006.

Heinecke, Liz Lee. *Radiant: The Dancer, the Scientist, and a Friendship Forged in Light*. New York: Grand Central Publishing, 2021.

Hope, Howard. *The Remarkable Life of Colonel George Edward Gouraud: His Origins; His Life and Times; His Associates*. Peterborough: Privately printed, 2021.

Hughes, Ivor, and David Ellis Evans. *Before We Went Wireless: David Edward Hughes, FRS: His Life, Inventions, and Discoveries*. Bennington, V: Images from the Past, 2011.

Insull, Samuel, and Larry Plachno. *The Memoirs of Samuel Insull*. Polo, IL: Transportation Trails, 1992.

Klein, Maury. *The Life and Legend of Jay Gould*. Baltimore: Johns Hopkins University Press, 1986.

Lewis, David I. *The Public Image of Henry Ford: An American Folk Hero and His Company*. Detroit: Wayne State University Press, 1976.

Marsh, J. O., and R. G. Roberts. "David Edward Hughes: Inventor, Engineer and Scientist." *Proceedings of the Institute of Electrical Engineers* 126 (1979): 929–35.

McDonald, Forrest. *Insull*. Chicago: University of Chicago Press, 1962.

McPartland, Donald S. "Almost Edison: How William Sawyer and Others Lost the Race to Electrification." PhD diss., City University of New York, 2006.

Middleton, William D., and William D. Middleton III. *Frank Julian Sprague: Electrical Inventor and Engineer*. Bloomington: Indiana University Press, 2009.

Morus. Iwan Rhys. *Nikola Tesla and the Electrical Future*. London: Icon Books, 2019.

O'Neill, John J. *Prodigal Genius: The Life of Nikola Tesla*. New York: I. Washburn, 1944.

Passer, Harold C. *Frank Julian Sprague, Father of Electric Traction, 1857–1934*. Cambridge, MA: Harvard University Press, 1952

Payne, Daniel G., ed. *Writing the Land: John Burroughs and His Legacy; Essays from the John Burroughs Nature Writing Conference*. Newcastle: Cambridge Scholars Publishing, 2008.

Renehan, Edward J., Jr. *John Burroughs: An American Naturalist*. Post Mills, VT: Chelsea Green Publishing, 1992.

Schneider, Janet M., and Bayla Singer, eds. *Blueprint for Change: The Life and Times of Lewis H. Latimer*. New York: Queens Borough Public Library, 1995.

Seifer, Marc J. *Wizard: The Life and Times of Nikola Tesla: Biography of a Genius*. New York: Citadel Press/Kensington Publishing, 1998.

Spehr, Paul C. *The Man Who Made Movies: W. K. L. Dickson*. New Barnet Herts., U.K.: John Libbey, 2008.

Sprague, Harriet Chapman Jones. *Frank J. Sprague and the Edison Myth*. New York: William-Frederick Press, 1947.

Sprague, John L. "Frank J. Sprague Invents: The Constant-Speed DC Electric Motor." *IEEE Power and Energy Magazine* 14 (2016): 80–96.

———. "A Sprague Invention: Multiple Unit Train Control." *IEEE Power and Energy Magazine* 13 (2015): 88–103.

Sprague, John L., and Joseph J. Cunningham. "A Frank Sprague Triumph: The Electrification of Grand Central Terminal." *IEEE Power and Energy Magazine* 11 (2013): 58–76.

Strouse, Jean. *Morgan: American Financier*. New York: Random House, 1999.

Szymanowitz, Raymond. *Edward Goodrich Acheson: Inventor, Scientist, Industrialist; A Biography*. New York: Vantage Press, 1971.

Tesla, Nikola. *My Inventions: The Autobiography of Nikola Tesla*. Radford, VA: Wilder Publications, 2007. (Orig. pub. 1919.)

Twain, Mark. *The Autobiography of Mark Twain*. New York: Harper & Bros., 1917.

Wasik, John F. *The Merchant of Power: Samuel Insull, Thomas Edison, and the Creation of the Modern Metropolis*. New York: Palgrave Macmillan, 2006.

Watts, Steven. *The People's Tycoon: Henry Ford and the American Century*. New York: Vintage Books, 2006.

Welch, Walter L. *Charles Batchelor: Edison's Chief Partner*. Syracuse, NY: Syracuse University Press, 1972.

Wile, Frederic William. *Emile Berliner: Maker of the Microphone*. Indianapolis: Bobbs-Merrill, 1926.

Wise, George. "Swan's Way: A Study in Style." *IEEE Spectrum* 19 (1982): 66–70.

Woodbury, David O. *Elihu Thomson: Beloved Scientist, 1853–1937*. Cambridge, MA: Harvard University Press, 1960.

———. *A Measure for Greatness: A Short Biography of Edward Weston*. New York: McGraw-Hill, 1949.

Worth, David S. *Spencer Trask: Enigmatic Titan*. New York: Kabique, Inc., 2008.

Index

About the Author and Contributors

Paul Israel is a historian of technology and the leading authority on Thomas Edison. He is the author of *Edison: A Life of Invention* (Wiley, 1998), *From Machine Shop to Industrial Laboratory: Telegraphy and the Changing Context of American Invention, 1830–1920* (Johns Hopkins University Press, 1992), and (with Robert Friedel and Bernard Finn) *Edison's Electric Light: The Art of Invention* (Johns Hopkins University Press, 2010) as well as numerous articles. He has been the director and general editor of the Thomas Edison Papers since 2002.

* * *

Ed Bradley, a consulting editor with the Thomas Edison Papers and an assistant editor with the Papers of Roger Brooke Taney, received his PhD in American history from the University of Illinois at Urbana-Champaign. He is the author of *"We Never Retreat": Filibustering Expeditions into Spanish Texas, 1812–1822*

(Texas A&M University Press, 2015). He lives in Pensacola, Florida.

Louis Carlat is an associate editor of the Thomas Edison Papers, where he has contributed to seven volumes. He received his PhD in the history of science, technology, and medicine from Johns Hopkins University.

Alexandra Rimer, M.L.S., is an assistant editor of the Thomas Edison Papers, where she has contributed to five volumes. She is the author of *Seduced by the Light: The Mina Miller Edison Story* (Lyon Press, 2023).

Daniel Weeks, PhD, is an associate research professor and editor at the Thomas Edison Papers. He is the author of *Gateways to Empire: Quebec and New Amsterdam to 1664* (Lehigh University Press, 2019) and *Not for Filthy Lucre's Sake: Richard Saltar and the Antiproprietary Movement in East New Jersey, 1664–1707* (Lehigh University Press, 2001).

Milton Keynes UK
Ingram Content Group UK Ltd.
UKHW051916240424
441696UK00005B/20